A Key to
Pacific Grasses

A Key to
Pacific Grasses

W. D. Clayton and Neil Snow

Kew Publishing
Royal Botanic Gardens, Kew

PLANTS PEOPLE
POSSIBILITIES

First published in 2010 by
Royal Botanic Gardens, Kew,
Richmond, Surrey, TW9 3AB, UK
www.kew.org

ISBN 978-1-84246-379-6

British Library Cataloguing in Publication Data
A catalogue record for this book is available from the British Library

Production editor: Ruth Linklater
Typesetting, page layout and cover design: Christine Beard
Publishing, Design & Photography, Royal Botanic Gardens, Kew

Front cover: *Heteropogon contortus* (pili grass), which was used extensively in the construction of grass huts in Hawaii and parts of the Pacific. Photo by Hank Oppenheimer (used with permission).

Printed by Lightening Source

For information or to purchase all Kew titles please visit
www.kewbooks.com or email publishing@kew.org

Kew's mission is to inspire and deliver science-based plant conservation worldwide, enhancing the quality of life.
All proceeds go to support Kew's work in saving the world's plants for life.

Contents

Introduction

- Materials and Methods 1
- Numerical summary and geographic origins 5

References 8

Acknowledgements 9

Key to Tribes 10

Key to Genera 12

Enumeration of species 23

Index 95

Introduction

The Pacific Ocean is the most expansive geographical feature on Earth. Included in its domain are thousands of atolls, smaller islands and, depending on how its boundaries are defined, several larger islands and island groups (Motteler 2006). Members of the grass family, Poaceae, are almost ubiquitous and are widespread across the Pacific.

The Philippines and New Guinea are frequently excluded from the Pacific based on biogeographical affinities, as in this treatment, and new grass Floras are badly needed for these areas. Modern treatments of Poaceae are available for New Zealand (Edgar & Connor 2000) and Australia (Sharp & Simon 2002), which we are also excluding from this treatment, despite the fact that New Zealand falls squarely into the realm of the Pacific based on its biogeography and its Polynesian heritage.

Approximately twenty years ago it became increasingly evident to workers in Hawaii, and those interested in the floristic diversity of the Pacific, that an updated taxonomic treatment of grasses for the Pacific region was needed. Funding from the National Science Foundation allowed the senior author to visit the Bernice P. Bishop Museum over an extended period to annotate its entire collection of Hawaiian and Pacific grasses. The purpose of this paper is to provide an updated taxonomic summary of the grasses from the Pacific.

Materials and Methods

Specimens of Poaceae from Hawaii and the Pacific have been examined at the Royal Botanic Gardens, Kew (K) and the Bernice P. Bishop Museum (BISH). Selected specimens were studied at a few other institutions although their holdings were not examined completely.

The taxonomic treatment at generic level is conservative and largely follows Clayton & Renvoize (1986). Thus, genera such as *Panicum* and *Stipa*, which are being segregated into smaller genera (e.g. Romaschenko *et al.* 2007; Zuloaga *et al.* 2007; Morrone *et al.* 2008), are retained in their broader sense. Users should be aware that specimens may be filed under different generic names in many herbaria and that different names are used in various online sources. The synonymy indicated for each species is rarely comprehensive but is intended to reflect names used most commonly in the Pacific. In some cases additional information is indicated when the correct application of a name is uncertain.

Many checklists and Floras retain the names of taxa for which one or only a few reports were ever noted, often many decades ago. Such taxa typically are non-native, appeared briefly, and were later extirpated. Because including them in summaries over-inflates estimates of biodiversity, taxa that were collected only once long ago, or which have not been collected in over fifty years, have been omitted. Likewise, species that were collected only in or near experimental trial plots, of which there have been many in Hawaii, are excluded.

To assemble data for this study it was necessary to rely on many older taxonomic works, which are indicated for many species. The bibliographic abbreviations include:

Fl. Poly.	Brown, F. B. H. (1931). Flora of southeastern Polynesia 1: Monocotyledons. *Bernice P. Bishop Mus. Bull.* 84.
List Micro.	Fosberg, F. R., Sachet, M.-H. & Oliver, R. (1987). Geographical checklist of Micronesia Monocotyledonae. *Micronesica* 20.
Fl. N. Cal.	Guillaumin, A. (1948). *Flore de la Nouvelle-Caledonie*. Paris.
Fl. Sol.	Hancock, I. R. & Henderson, C. P. (1988). *Flora of the Solomon Is.* Research Bull. No. 7. Honiara.
Pl. Samoa	Parham, B. E. V. (1972). Plants of Samoa. *New Zealand Dept. Sci. Industr. Res. Inform. Ser.* 85.
Fl. Fiji	Smith, A. C. (1979). *Flora Vitiensis* Vol. 1. Pacific Tropical Botanical Garden, Lawai, Hawaii.
Fl. Fann.	St. John, H. (1974). The vascular flora of Fanning Is., Line Ils. *Pacific Sci.* 28.
Fl. Pit.	St. John, H. (1987). An account of the flora of Pitcairn Island. *Pacific Plant Studies*, Honolulu 46.
Fl. Guam	Stone, B. C. (1970). Flora of Guam. *Micronesica* 6.
Fl. Niue	Sykes, W. R. (1970). Contribution to the Flora of Niue. *Bull. New Zealand Dept. Sci. Industr. Res.* 200.
Man. Haw.	Wagner, W. L., Herbst, D. R. & Sohmer, S. H. (1990). *Manual of the flowering plants of Hawaii*, Vol. 2. Bishop Museum Press, Honolulu.
Fl. Soc.	Welsh, S. L. (1998). *Flora Societatis*. Brigham Young University Press.
Fl. Raro.	Wilder, G. P. (1931). Flora of Rarotonga. *Bernice P. Bishop Mus. Bull.* 86.
Pl. Tonga	Yuncker, T. G. (1959). Plants of Tonga. *Bernice P. Bishop Mus. Bull.* 220.
Easter	Zizka, G. (1991). Flowering plants of Easter Island. *Palm. Hortus Francofurt.* 3.

(Other abbreviations in the taxonomic account follow Stafleu & Cowan (1976) and its supplements)

Many grass species in the Pacific are non-native (see below). Since many non-natives potentially can grow along roadsides or in other moderately disturbed and relatively open sites, comments on habitat occurrence are only given when a species has strong ecological preferences or is known from only one habitat type from the Pacific.

For geographic abbreviations we follow the three-letter geographic acronyms of Brummitt (2001) (Table 2, p. 7) and as recommended by the Taxonomic Databases Working Group (now known as Biodiversity Information Standards, http://www.tdwg.org/).

Bamboos. Over 1300 woody bamboo species exist worldwide (Judziewicz *et al.* 1999) and many more are undescribed. No attempt is made here to document carefully the distribution of woody bamboos in the Pacific, given that so many taxa are planted locally for various purposes (e.g. Staples & Herbst 2005). It has been estimated recently that some 200 bamboo species are in cultivation on the mainland in the United States (L. Clark, pers. comm., 2008), and we suspect at least two dozen species are in cultivation in the Pacific. The frequent persistence of bamboos after cultivation along or near abandoned fields, their infrequent flowering, and the considerable difficulty associated with identifying sterile material were among the reasons for excluding them from the detailed account.

Native bamboos in the Pacific include *Bambusa solomonensis* Holttum (endemic to CRL, SOL), two endemic species of *Greslania* in New Caledonia (*G. circinnata* Balansa, *G. rivularis* Balansa), and *Schizostachyum tessellatum* A. Camus (SOL). Some commonly planted bamboos in the Pacific include (see also Staples & Herbst 2005): *Bambusa balcooa* Roxb. (HAW), *B. arundinacea* (Retz.) Willd. (HAW), *B. beecheyana* Munro (HAW), *B. blumeana* Schult. f. (MRN), *B. multiplex* Raeusch. (CRL, FIJ, HAW, MRN, SCI, SOL), *B. oldhamii* Munro (HAW), *B. tuldoides* Munro (HAW), *B. vulgaris* Schrad. (CRL, FIJ, HAW, SOL); *Nastus elatus* Holttum (VAN), *N. obtusus* Holttum (SOL), *N. productus* (Pilg.) Holttum (SOL); *Otatea acuminata* (Munro) C. E. Calderón & Soderstr. (HAW), *O. aztecorum* (McClure & E. W. Sm.) C. E. Calderón & Soderstr. (HAW); *Phyllostachys aurea* Carrière ex Rivière & C. Rivière (HAW), *P. nigra* (Lodd. ex Lindl.) Munro (HAW); *Racemobambos holttumii* S. Dransf. (SOL); and *Schizostachyum glaucifolium* (Rupr.) Munro (FIJ, HAW, MRQ, SCI, SOL).

Invasive species. The spread of invasive, non-native taxa, and their impacts on native vegetation, has accelerated greatly in recent decades (Meyer 2000). Many invasive species can become weedy along roadsides or other periodically disturbed, relatively mesic sites. However, some grasses are among the most aggressive invasive terrestrial plant species in the Pacific. In Hawaii the worst invaders include *Pennisetum setaceum*, which may grow in stands dense enough to carry fire, and *Panicum maximum*, which can be so dense that it crowds out virtually all other herbaceous species. Other grasses that are highly invasive in some areas of the Pacific (Meyer 2000) include *Heteropogon contortus*, *Imperata cylindrica*, *Melinis minutiflora*, *Panicum repens*, *Paspalum conjugatum*, *P. distichum*, *Pennisetum clandestinum*, *P. polystachion*, *P. purpureum*, *Sorghum halepense* and *S.* ×*drummondii*.

Threatened species. Walter & Gillett (1998) list 29 species as being under varying degrees of threat.

Rare: *Leptochloa marquisensis*, *Panicum isachnoides*, *Pennisetum marquisense* (= *Cenchrus caliculatus*).

Vulnerable: *Calamagrostis expansa*, *Dissochondrus biflorus*, *Festuca hawaiensis*, *Ischaemum byrone*, *Leptochloa xerophila*, *Panicum koolauense*, *P. ramosius*, *Setaria jaffrei*.

Endangered: *Ancistrachne numaeensis*, *Calamagrostis hillebrandii*, *Cenchrus agrimonioides*, *Chloris cheesemanii* (= *Enteropogon unispiceus*), *Eragrostis fosbergii*, *Garnotia cheesemanii*, *Ischaemum polystachyum*, *Oryza neocaledonica*, *Panicum beecheyi*, *P. fauriei*, *P. lineale*, *P. niihauense*, *Poa sandvicensis*, *P. siphonoglossa*, *Trisetum inaequale*.

Endangered or extinct: *Eragrostis hosakai* (= *E. leptostachya*), *E. mauiensis*, *Stipa horridula* (= *S. scabra*).

Of these, *Chloris cheesemanii*, *Eragrostis hosakai*, *Pennisetum marquisense* and *Stipa horridula* are treated here as synonymous with more widespread species which are not at risk. Even so, the remainder amount to a quarter of the endemic grass flora.

Tribal key. A key to grass tribes is included, although tribal boundaries remain somewhat in taxonomic flux. While the tribal key does not attempt to account for aberrations from the most typical situation for each genus, it should work in most instances and hopefully the tabular summary (Table 1) will help users become better acquainted with the tribes.

TABLE 1. An alphabetical tabular summary of tribes indicating the generic composition of each.

Andropogoneae (*Andropogon*, *Apluda*, *Arthraxon*, *Bothriochloa*, *Capillipedium*, *Chionachne*, *Chrysopogon*, *Coix*, *Cymbopogon*, *Dichanthium*, *Dimeria*, *Elionurus*, *Eremochloa*, *Eulalia*, *Hackelochloa*, *Hemarthria*, *Heteropogon*, *Hyparrhenia*, *Imperata*, *Ischaemum*, *Microstegium*, *Miscanthus*, *Mnesithea*, *Pogonatherum*, *Polytrias*, *Rottboellia*, *Saccharum*, *Schizachyrium*, *Sorghum*, *Themeda*, *Tripsacum*, *Zea*)

Aristideae (*Aristida*)

Arundineae (*Arundo*, *Cortaderia*, *Phragmites*, *Rytidosperma*)

Arundinelleae (*Garnotia*)

Aveneae (*Aira*, *Agrostis*, *Alopecurus*, *Ammophila*, *Anthoxanthum*, *Arrhenatherum*, *Avena*, *Calamagrostis*, *Deschampsia*, *Dichelachne*, *Gastridium*, *Holcus*, *Koeleria*, *Phalaris*, *Phleum*, *Polypogon*, *Trisetum*)

Bambuseae (*Bambusa*, *Greslania*, *Nastus*, *Otatea*, *Phyllostachys*, *Racemobambos*, *Schizostachyum*)

Bromeae (*Bromus*)

Centotheceae (*Centotheca*, *Lophatherum*)

Cynodonteae (*Chloris*, *Cynodon*, *Enteropogon*, *Eustachys*, *Lepturopetium*, *Lepturus*, *Tragus*, *Zoysia*)

Ehrharteae (*Ehrharta*)

Eragrostideae (*Dactyloctenium*, *Distichlis*, *Ectrosia*, *Ectrosiopsis*, *Eleusine*, *Eragrostis*, *Leptochloa*, *Muhlenbergia*, *Sporobolus*)

Eriachneae (*Eriachne*)

Isachneae (*Isachne*)

Meliceae (*Glyceria*)

Oryzeae (*Oryza*, *Zizania*)

Paniceae (*Alloteropsis*, *Ancistrachne*, *Anthephora*, *Axonopus*, *Cenchrus*, *Cyrtococcum*, *Digitaria*, *Dissochondrus*, *Echinochloa*, *Entolasia*, *Eriochloa*, *Ixophorus*, *Melinis*, *Oplismenus*, *Panicum*, *Paspalidium*, *Paspalum*, *Pennisetum*, *Sacciolepis*, *Setaria*, *Spinifex*, *Stenotaphrum*, *Thuarea*, *Urochloa*)

Phareae (*Leptaspis*, *Scrotochloa*)

Poeae (*Briza*, *Dactylis*, *Festuca*, *Lamarckia*, *Lolium*, *Poa*, *Vulpia*)

Stipeae (*Oryzopsis*, *Stipa*)

Thysanolaeneae (*Thysanolaena*)

Triticeae (*Elymus*, *Hordeum*, *Triticum*)

Numerical summary and geographic origins

This study treats 420 species of herbaceous grasses in 120 genera. It does not account for infraspecific taxa, which would have required time resources beyond what was available. As a general statement, the Pacific region is not particularly diverse for native grasses. While the study summarises knowledge to date, species of grasses are being newly reported from the Pacific at steady rates. Most attention has been devoted to the grass flora of Hawaii (e.g. Herbst & Clayton 1998; Oppenheimer 2008; Snow 2008), but many additional reports are to be expected from throughout the Pacific. Furthermore, we wish to stress that many native grass species in the Pacific would benefit from additional taxonomic study.

Floras in the Pacific islands are derived from natural dispersal, Polynesian and European introductions, and autochthonous or vicariant origins. Disentangling these dispersal modes is not a simple matter and it is beyond our scope to delve into a critical analysis; however, we can attempt an approximation. The following figures omit crop species but include endemic and naturalised bamboos.

Disjunct		
Europe & Middle East	54 species	
Africa	33	
America	52	Total 139
Tropical		
Pantropical	57	
Palaeotropical	26	
Southeast Asia, some extending to India or Australia	85	Total 168
Australasia	33	Total 33
Pacific		
Melanesia, some extending to New Guinea	34	
Micronesia	10	
Polynesia	13	
Hawaii	37	
Wider Pacific distribution	5	Total 99
		Grand total 439

The disjunct species were evidently introduced subsequent to the European incursion into the Pacific. The dispersal of Tropical and Australasian species is less clear, but we may make a tentative assumption that species reaching islands nearest the mainland source were probably capable of dispersing naturally or during the Polynesian diaspora (we can see no way to distinguish between these), whereas those occurring only on more distant islands are likely to be post-European. By this criterion 77 of the Tropical and 16 of the Australasian species can be ascribed to European introduction.

Bringing these figures together we can apportion the origin of the Pacific grass flora to 23% endemic, 24% by natural or Polynesian dispersal from Asia or Australia and 53% introduced subsequent to the European incursion. This calculation is clearly speculative, but it affords a reasonable estimate of how the Pacific acquired its grass flora.

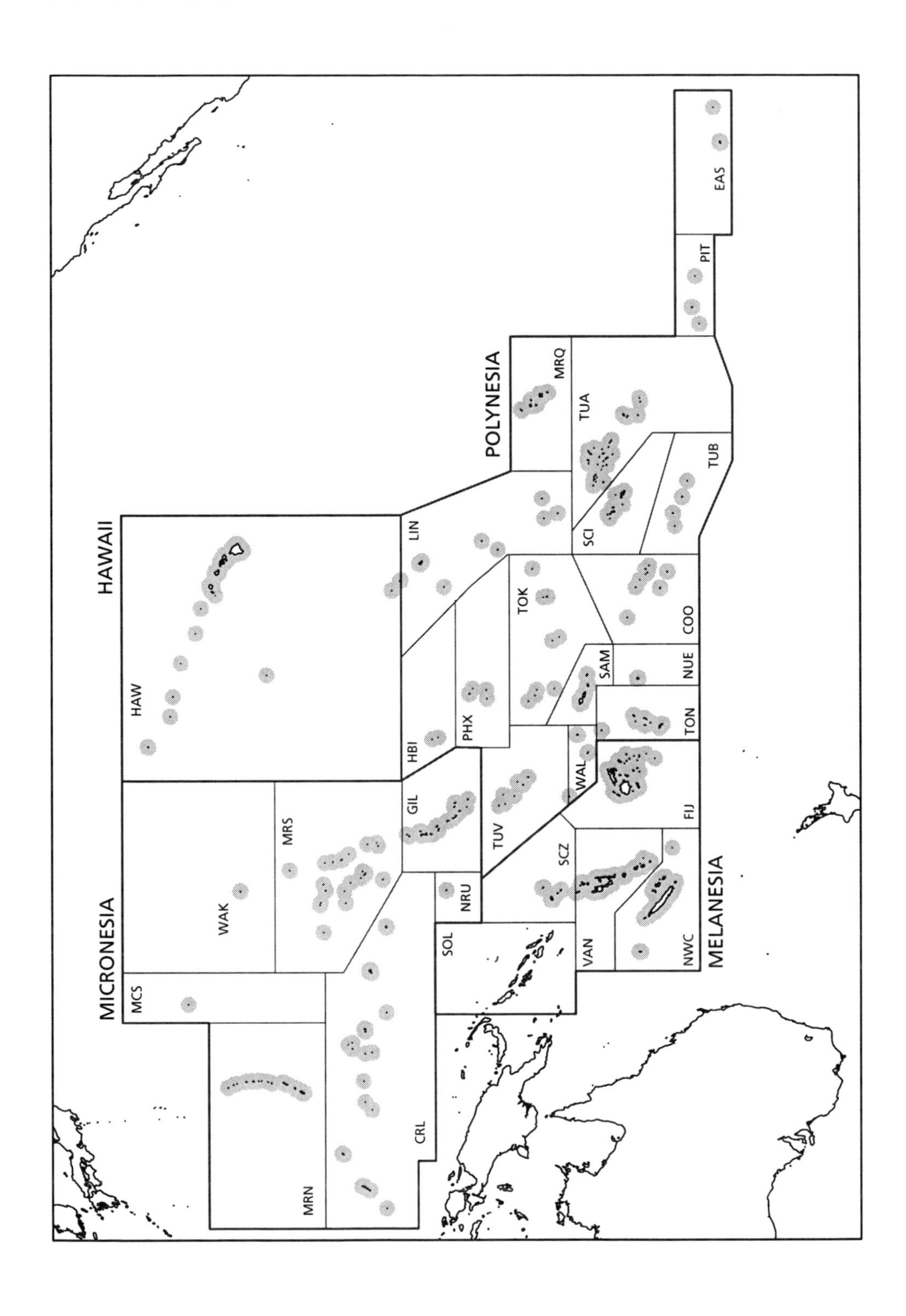
HAWAII
POLYNESIA
MICRONESIA
MELANESIA
HAW
LIN
MRQ
TUA
TUB
EAS
PIT
SCI
TOK
COO
SAM
NUE
TON
PHX
HBI
GIL
MRS
WAK
MCS
MRN
CRL
NRU
TUV
WAL
FIJ
SCZ
VAN
NWC
SOL

TABLE 2. Alphabetical listing of acronyms used for geographic distributions following Brummitt (2001). Another authoritative and recent guide to names used in the Pacific is Motteler (2006). See also Map on facing page.

COO	Cook Is.	PHX	Phoenix Is.
CRL	Caroline Is.	PIT	Pitcairn Is.
EAS	Easter Is.	SAM	Samoa
FIJ	Fiji	SCI	Society Is.
GIL	Gilbert Is.	SCZ	Santa Cruz Is.
HAW	Hawaii	SOL	Solomon Is.
HBI	Howland-Baker Is.	TOK	Tokelau-Manihiki
LIN	Line Is.	TON	Tonga
MCS	Marcus Is.	TUA	Tuamoto
MRN	Marianas	TUB	Tubuai Is.
MRQ	Marquesas	TUV	Tuvalu
MRS	Marshall Is.	VAN	Vanuatu
NRU	Nauru	WAK	Wake Is.
NUE	Niue	WAL	Wallis-Futuna Is.
NWC	New Caledonia		

References

Brummitt, R. K. (2001). *World Geographical Scheme for Recording Plant Distributions*. Edition 2. Hunt Institute for Botanical Documentation, Pittsburgh.

Clayton, W. D. & Renvoize, S. A. (1986). Genera Graminum: Grasses of the World. *Kew Bull. Addit. Ser.* XIII. Her Majesty's Stationery Office, London.

Edgar, E. & Connor, H. E. (2000). *Flora of New Zealand*, Vol. 5. Gramineae. Manaaki Whenua Press, Lincoln, New Zealand.

Herbst, D. R. & Clayton, W. D. (1998). Notes on the grasses of Hawai'i: new records, corrections, and name changes. *Occas. Pap. Bernice Puauhi Bishop Mus.* 55: 17–38.

Judziewicz, E. J., Clark, L. G., Londoño, X. & Stern, M. J. (1999). *American Bamboos*. Smithsonian Press, Washington, D.C.

Meyer, J.-Y. (2000). Preliminary review of the invasive plants in the Pacific islands (SPREP Member Countries), pp. 85–114. In: G. Sherley (ed.), *Invasive species in the Pacific: A technical review and draft regional strategy*. South Regional Environment Programme, Samoa.

Morrone, O., Denham, S. S., Aliscioni, S. S. & Zuloaga, F. O. (2008). *Parodiophyllochloa*, a new genus segregated from *Panicum* (Paniceae, Poaceae) based on morphological and molecular data. *Syst. Bot.* 33: 66–76.

Motteler, L. S. (2006). *Pacific Island names: A map and name guide to the New Pacific*. Bishop Museum Press, Honolulu.

Oppenheimer, H. (2008). New Hawaiian plant records for 2007. *Occas. Pap. Bernice Puauhi Bishop Mus.* 100: 22–38.

Romaschenko, K., Petersen, P. M., Garcia-Jacas, N., Soreng, R. J. & Alfonso, S. (2007). A phylogeny of *Stipeae* (Poaceae) based on nuclear DNA (ITS) sequence data. *Abstract 2734, BOTANY 2007*. Chicago, IL.

Sharp, D. & Simon, B. K. (2002). *AusGrass*: *Grasses of Australia*. CD-ROM plus users guide. CSIRO Publishing, Collingwood, VIC, Australia.

Snow, N. (2008). Notes on grasses (Poaceae) in Hawai'i. *Occas. Pap. Bernice Puauhi Bishop Mus.* 100: 38–43.

Stafleu, F. A. & Cowan, R. S. (1976–2009). *Taxonomic Literature* Vols. 1–7 and Supplements 1–7. Bohn, Scheltema & Holkema, Utrecht.

Staples, G. W. & Herbst, D. R. (2005). *A Tropical Garden Flora*. Bishop Museum Press, Honolulu.

Walter, K. S. & Gillett, H. J. (1998). *1997 IUCN Red Book of Threatened Plants*. IUCN.

Zuloaga, F. O., Giussani, L. M. & Morrone, O. (2007). *Hopia*, a new monotypic genus segregated from *Panicum* (Poaceae). *Taxon* 56: 145–156.

Acknowledgements

Many people have answered questions or otherwise helped in this project over its long duration, including Danielle Frohlich, Napua Harbottle, Derral Herbst, Clyde Imada, Surrey Jacobs, Barbara Kennedy, Walter Kittredge, Alex Lau, David Lorence, Bryan Simon, Paul Petersen, Jef Veldkamp, Warren L. Wagner and George Yatskievych, not forgetting our meticulous editor Tom Cope. We ask forgiveness from those whose names we may have unintentionally omitted. We also thank the larger cadre of botanists who have collected plants in the Pacific and whose interesting discoveries sometimes go unreported for too long. This project was partially supported by the National Science Foundation (DEB-8912364 to S. H. Sohmer) and the Marks Endowment at the Bishop Museum. This paper represents Contribution number 2008-039 from the Pacific Biological Survey.

Key to Tribes

1. Spikelets 1–many-flowered, if 2-flowered then both bisexual or the upper imperfect:
   2. Leaf-blades with distinct cross-veins beneath, usually lanceolate or elliptic:
      3. Leaf venation slanting obliquely from midrib; lemma margins closed except for an apical pore; spikelets unisexual .... **Phareae**
      3 Leaf venation parallel; lemma margins free; spikelets bisexual .... **Centotheceae**
   2. Leaf-blades without cross-veins:
      4. Glumes absent; spikelets 1-flowered .... **Oryzeae**
      4. Glumes, or at least the upper, present though sometimes much reduced:
         5. Margins of lemma not clasping palea keels, sometimes enfolding palea but then spikelets 1-flowered, both scales usually membranous:
            6. Spikelets with 2 sterile florets below the single fertile floret and falling with it; inflorescence a loose panicle or raceme .... **Ehrharteae**
            6. Spikelets usually without sterile florets below the fertile, if present then inflorescence a spiciform or capitate panicle:
               7. Inflorescence a large plumose panicle; spikelets several-flowered .... **Arundineae**
               7. Inflorescence not a plumose panicle:
                  8. Lemma 5- or more-veined (3-veined in *Dichelachne crinita*, *Koeleria macrantha*, *Oryzopsis miliacea*):
                     9. Inflorescence a single bilateral raceme, the spikelets inserted broadside on opposite sides of the rhachis .... **Triticeae**
                     9. Inflorescence a panicle, rarely a raceme and then spikelets inserted edgeways on:
                        10. Ovary capped by a hairy lobed appendage, the stigmas subterminal; lemma usually with a subterminal awn; leaf-sheaths usually hairy .... **Bromeae**
                        10. Ovary glabrous or hairy but not appendaged, the stigmas terminal:
                            11. Glumes shorter than the spikelet (longer in *Lolium temulentum* which has a bilateral raceme); spikelets dull:
                                12. Upper glume 1-veined, obtuse (as to the typical genus *Glyceria*) .... **Meliceae**
                                12. Upper glume 3–9-veined, rarely 1-veined and then acuminate; spikelets 2–many-flowered (except *Lamarckia*) .... **Poeae**
                            11. Glumes equalling or exceeding the rest of the spikelet; rarely shorter and then spikelets shiny (if accompanied by sterile spikelets see Poeae — *Lamarckia*):
                                13. Ligule membranous:

14. Floret laterally compressed, usually membranous, with or without a dorsal awn; spikelets usually shiny, 1–several-flowered .... **Aveneae**
14. Floret terete or dorsally compressed, more or less indurate, with a terminal awn; spikelets 1-flowered .... **Stipeae**
13. Ligule a line of hairs; lemmas awned from the sinus of a bilobed apex .... **Arundineae**
8. Lemma 1–3-veined (*Distichlis* 5–11-veined):
15. Fertile lemma 3-awned .... **Aristideae**
15. Fertile lemma with or without a single awn:
16. Inflorescence a panicle, or composed of racemes and then spikelets several-flowered:
17. Spikelets breaking up at maturity (or with several fertile florets — *Eragrostis superba*) .... **Eragrostideae**
17. Spikelets falling entire, 1-flowered (as to the atypical genus *Garnotia*) .... **Arundinelleae**
16. Inflorescence composed of racemes; spikelets with 1 fertile floret .... **Cynodonteae**
5. Margins of lemma clasping palea keels, both scales usually indurated; spikelets strictly 2-flowered (if subtended by 1 or more bristles see Paniceae — *Dissochondrus*):
18. Spikelets dorsally compressed; lemma at most puberulous, never awned ... . **Isachneae**
18 Spikelets slightly laterally compressed; lemmas pilose or awned .... **Eriachneae**
1. Spikelets 2-flowered, the lower male or barren, the upper bisexual:
19. Spikelets tardily breaking up at maturity (if lower lemma dorsally awned see Aveneae – *Arrhenatherum*) .... **Isachneae**
19. Spikelets falling entire at maturity:
20. Spikelets falling with pedicel attached, tiny, borne in a large panicle ....**.Thysanolaeneae**
20. Spikelets falling singly without pedicel, or falling in pairs or clusters:
21. Glumes thinner than upper lemma, this more or less indurated; lower glume often shorter than spikelet .... **Paniceae**
21. Glumes thicker than upper lemma, this hyaline to membranous; lower glume as long as spikelet or almost so .... **Andropogoneae**

Key to Genera

1. Spikelets 1–many-flowered, if 2-flowered then both bisexual or the upper imperfect:
    2. Leaf-blades with distinct cross-veins beneath, usually lanceolate to elliptic:
        3. Lemma margins closed except for an apical pore; leaf-blade venation slanting obliquely from midrib; spikelets unisexual, the male small:
            4. Female spikelets symmetrical, urn-like, with terminal pore; panicle branches whorled, stiff .... **1. Scrotochloa**
            4. Female spikelets asymmetrical, conchiform, with lateral pore .... **2. Leptaspis**
        3. Lemma margins free; leaf-blade venation parallel; spikelets bisexual, falling entire:
            5. Spikelets awnless; upper lemmas developing reflexed bristles at maturity .... **37. Centotheca**
            5. Spikelets terminating in a fan of awns, these hooked at maturity .... **38. Lophatherum**
    2. Leaf-blades without cross-veins:
        6. Spikelets with 2 or more fertile florets:
            7. Inflorescence a panicle, open, contracted or spiciform (rarely the branches indistinctly racemose, but then compacted into an oblong head — *Distichlis*):
                8. Spikelets laterally compressed; lemmas membranous to coriaceous:
                    9. Lemmas bearing tufts of hair, deeply bilobed, with a geniculate awn from the sinus .... **39. Rytidosperma**
                    9. Lemmas glabrous or hairy but not in tufts; awn, if present, apical, subapical or dorsal:
                        10. Panicle plumose, large, on robust culms:
                            11. Upper glume 1-veined; plants gynodioeceous (female and bisexual) .... **40. Cortaderia**
                            11. Upper glume 3–5-veined; plants all bisexual:
                                12. Lemmas plumose; florets all fertile .... **41. Arundo**
                                12. Lemmas glabrous, the hairs arising from callus and rhachilla; lowest floret male or barren .... **42. Phragmites**
                        10. Panicle not plumose:
                            13. Ligule membranous, its upper margin glabrous or obscurely ciliolate; lemma 5–13-veined (3-veined in *Koeleria*):
                                14. Lemmas broader than long, cordate, keeled below but rounding above as grain swells, with broad membranous margins; spikelets borne in trembling panicles .... **12. Briza**
                                14. Lemmas lanceolate to ovate, not cordate:
                                    15. Lemmas keeled:
                                        16. Basal shoots compressed, keeled and interleaved fanwise; panicle 1-sided, lobed, the spikelets in dense fascicles; lemma acuminate to briefly awned, stiffly stiffly ciliate on keel .... **14. Dactylis**

16. Basal shoots not fan-like; panicle not forming dense 1-sided fascicles; lemma keel glabrous or softly hairy:
  17. Panicle open, spikelets dull:
    18. Lemmas 2.5–5 mm long; awnless ..... **13. Poa**
    18. Lemmas 13–23 mm long; awns 1–9 mm long .. .. **33. Bromus**
  17. Panicle contracted or spiciform, though often lobed or interrupted, the axis pubescent; spikelets shiny:
    19. Lemmas awnless .... **18. Koeleria**
    19. Lemmas awned from back or below tip .... **19. Trisetum**

15. Lemmas rounded on the back:
  20. Glumes shorter than spikelet:
    21. Glumes obtuse, the upper 1-veined; lemma 7-veined, awnless ....**15. Glyceria**
    21. Glumes acute to acuminate, the upper 5–7-veined (sometimes 1-veined in *Vulpia*):
      22. Awn subapical (*B. inermis* awnless); leaf-sheaths hairy (*B. secalinus* usually glabrous); lemma 7-veined; ovary capped by a fleshy hairy appendage above insertion of stigmas .... **33. Bromus**
      22. Awn, when present, apical; leaf-sheaths glabrous (except sometimes in *Festuca rubra*); lemma 5-veined; ovary glabrous, or pubescent but then without appendage:
        23. Plants perennial; awns absent or up to 5 mm long .... **8. Festuca**
        23. Plants annual; awns 5–15 mm long .... **9. Vulpia**
  20. Glumes as long as spikelet and enclosing florets; awn dorsal (sometimes absent in *Avena sativa*):
    24. Spikelets 18–30 mm long, 2–3-flowered; ovary hairy .... **17. Avena**
    24. Spikelets 2.5–6 mm long, strictly 2-flowered; ovary glabrous:
      25. Plants perennial; florets separated by an internode and surmounted by a rhachilla extension .... **20. Deschampsia**
      25. Plants annual; florets without internodes or rhachilla extension .... **21. Aira**

13. Ligule represented by a fringe of hairs (surmounting a brief membrane in *Distichlis*):
  26. Glumes as long as spikelet, enclosing 2 florets, these both fertile and awned, without rhachilla extension ..... **87. Eriachne**
  26. Glumes shorter than spikelet:

27. Lemma 5–11-veined; leaves distichous, stiff with a hard tip; inflorescence a compact head of 1–several brief racemose branches, but these sometimes reduced to only 2–3 spikelets; dioecious .... **45. Distichlis**

27. Lemma 3-veined; leaves not distichous:

28. Florets awnless .... **46. Eragrostis**

28. Florets awned:

29. Fertile florets 10–25, subequal, disarticulating between them at maturity; awns 0.5–1.5 mm long .... **47. Ectrosiopsis**

29. Fertile florets 1–2, with longer sterile florets above, disarticulating above glumes but not between florets; awns 3–8 mm long .... **48. Ectrosia**

8. Spikelets dorsally compressed, 2-flowered, awnless; lemmas coriaceous to crustaceous:

30. Inflorescence spiciform, the spikelets subtended by 1 or more sterile bristles .... **78. Dissochondrus**

30. Inflorescence an open or contracted panicle, without sterile bristles .... **86. Isachne**

7. Inflorescence composed of separate racemes bearing many spikelets:

31. Lower glume absent; spikelets attached edgeways on and partially sunk in axis of a single raceme .... **10. Lolium**

31. Lower glume present; spikelets attached broadside to rhachis and free from it:

32. Raceme bilateral, single, the spikelets alternate on opposite sides:

33. Plants perennial, with long rhizomes .... **34. Elymus**

33. Plants annual, cultivated .... **35. Triticum**

32. Racemes unilateral, usually several, the spikelets generally imbricate:

34. Racemes borne along a central axis .... **49. Leptochloa**

34. Racemes single, paired or digitate:

35. Fertile florets 2, awned .... **60. Lepturopetium**

35. Fertile florets 3–9, at most mucronate:

36. Rhachis terminating in a spikelet .... **50. Eleusine**

36. Rhachis terminating in a bare point ..... **51. Dactyloctenium**

6. Spikelets with 1 fertile floret, with or without additional sterile florets:

37. Spikelets with sterile lemmas below the fertile and deciduous with it, usually 2 (imitating glumes in *Oryza*), but sometimes 1 of them suppressed in *Phalaris*:

38. Glumes absent .... **3. Oryza**

38. Glumes present:

39. Glumes much shorter than spikelet, sometimes reduced to obscure lips (*E. stipoides*); panicle loose .... **5. Ehrharta**

39. Glumes, or at least the upper, as long as spikelet; panicle spiciform or capitate:

40. Sterile lemmas awned, exceeding the fertile ... **23. Anthoxanthum**

40. Sterile lemmas awnless, often much reduced .... **24. Phalaris**

37. Spikelets without sterile lemmas below the fertile:

41. Inflorescence an open, contracted or spiciform panicle:

42. Glumes absent; spikelets unisexual, the male in lower part of panicle .. .. **4. Zizania**
42. Glumes present; spikelets bisexual:
    43. Spikelets shedding florets above the more or less persistent glumes:
        44. Spikelets with a cluster of awned sterile lemmas above the fertile and falling with it .... **48. Ectrosia**
        44. Spikelets with or without a filiform rhachilla extension above the fertile lemma:
            45. Lemmas 3-awned .... **44. Aristida**
            45. Lemmas with or without a single awn (except 5 awns for *Calamagrostis hillebrandii*):
                46. Ligule reduced to a line of hairs; lemma 1-veined, awnless; pericarp peeling from seed when wetted .... **52. Sporobolus**
                46. Ligule membranous, this rarely with a ciliolate edge; lemma 3–7-veined; pericarp adherent to seed:
                    47. Glumes with a rounded swelling at the base, this housing the short floret; panicle spiciform .... **29. Gastridium**
                    47. Glumes not swollen at the base:
                        48. Floret callus pungent, 1.5–2 mm long; lemma terete, with a long terminal geniculate awn ... . **6. Stipa**
                        48. Floret callus obtuse, inconspicuous:
                            49. Glumes shorter than floret, or slightly longer but then gaping to expose it:
                                50. Lemma 3-veined, sometimes with a terminal awn; lower glume up to $^{3}/_{4}$ length of floret .... **53. Muhlenbergia**
                                50. Lemma 5-veined, with a subapical or dorsal awn; glumes about as long as floret ... . **19. Trisetum**
                            49. Glumes much exceeding and enclosing the floret:
                                51. Lemma dorsally compressed, with a deciduous terminal awn .. **7. Oryzopsis**
                                51. Lemma laterally compressed, with or without a dorsal or subapical awn:
                                    52. Glumes ciliate on keel, shortly awned, both 3-veined; panicle spiciform, cylindrical; lemma awnless .... **32. Phleum**
                                    52. Glumes at most scabrid on keel, awnless, the lower or both 1-veined:

53. Lemma with or without a long (8–40 mm) flexuous awn . . . . **28. Dichelachne**
53. Lemma with or without a short (0–6 mm) awn:
54. Leaf-blades flexible, acute; ligule 0.5–6 mm long; panicle open or contracted, rarely spiciform but then the lemma awned; spikelets 2–8 mm long:
55. Rhachilla not or scarcely prolonged beyond base of floret (except *A. avenacea* which has a hairy lemma); floret callus glabrous or almost so . . . . **25. Agrostis**
55. Rhachilla clearly prolonged as a plumose bristle; floret callus pubescent or bearded; lemma glabrous . . . . **26. Calamagrostis**
54. Leaf-blades stiff, pungent; ligule 10–30 mm long; panicle spiciform; spikelets 10–16 mm long; lemma mucronate . . . . **27. Ammophila**
43. Spikelets falling entire at maturity:
56. Spikelets in clusters of 1 fertile and 4 sterile, all falling together . . . . **11. Lamarckia**
56. Spikelets all fertile, falling singly:
57. Florets 2, the upper male with a hooked awn; foliage softly hairy . . . . **22. Holcus**
57. Floret 1:
58. Lemma with a dorsal awn; panicle spiciform, cylindrical . . . . **31. Alopecurus**
58. Lemma with or without a terminal awn; glumes almost as long as or exceeding the floret (if much shorter see *Oryza* in which sterile lemmas imitate glumes):
59. Spikelets disarticulating from top of pedicel; lemma 3-veined . . . . **88. Garnotia**
59. Spikelets disarticulating from base of pedicel, or (*P. monspeliensis*) from near its top but leaving an oblong callus at base of spikelet; lemma 5-veined . . . . **30. Polypogon**

41. Inflorescence composed of racemes:
60. Spikelets in groups of 3 on opposite sides of the rhachis . . . **36. Hordeum**
60. Spikelets single or paired:
61. Racemes unilateral, single, paired or digitate; spikelets breaking up:
62. Spikelets strictly 1-flowered, at most with a filiform rhachilla extension:
63. Rhachis tough, the spikelets free; racemes digitate . . . . **57. Cynodon**
63. Rhachis fragile, cylindrical, the spikelets sunk in hollows; raceme single . . . . **61. Lepturus**
62. Spikelets with sterile florets above the fertile:
64. Lemma and grain dorsally compressed . . . **56. Enteropogon**
64. Lemma and grain laterally compressed:
65. Upper glume 5–7-veined . . . . **60. Lepturopetium**
65. Upper glume 1(–3)-veined; racemes digitate:
66. Lemma awned; upper glume acute, awnless . . . . **54. Chloris**
66. Lemma awnless, golden brown; upper glume obtuse, briefly awned . . . . **55. Eustachys**
61. Raceme multilateral, single, like a bottle-brush; spikelets falling entire:
67. Upper glume spinosely ribbed; spikelets paired . . . . . **58. Tragus**
67. Upper glume smooth; spikelets single . . . . **59. Zoysia**
1. Spikelets 2-flowered, the lower male or barren, the upper fertile:
68. Spikelets breaking up at maturity:
69. Florets laterally compressed, the lower geniculately awned from the back (if awnless see *Phalaris*, in which 1 of the sterile florets is sometimes suppressed) . . . . **16. Arrhenatherum**
69. Florets dorsally compressed, awnless . . . . **86. Isachne**
68. Spikelets falling entire at maturity (or persisting on plant in crop species):
70. Spikelets falling singly with pedicel attached, 1–2 mm long, borne along the branches of a large panicle; glumes and lemmas membranous . . . . **43. Thysanolaena**
70. Spikelets falling singly without attached pedicel, or falling in pairs or clusters:
71. Glumes thinner than upper lemma, this more or less indurated; lower glume often shorter than spikelet:
72. Plant dioecious, the female inflorescence a large globose head of spiny quills each bearing a basal spikelet . . . . **85. Spinifex**
72. Plant bisexual:
73. Inflorescence a panicle, sometimes spiciform or irregularly contracted about primary branches:
74. Spikelets not subtended by bristles or scales:
75. Panicle spiciform, cylindrical; spikelets gibbous . . **65. Sacciolepis**
75. Panicle open or contracted:
76. Upper lemma cartilaginous with flat margins, laterally compressed; lower glume minute; upper glume emarginate to bilobed, sometimes mucronate or awned . . **80. Melinis**

76. Upper lemma coriaceous to crustaceous with inrolled margins; lower glume developed; upper glume entire, awnless:
77. Spikelets laterally compressed, gibbous; upper lemma with a little green crest .... **66. Cyrtococcum**
77. Spikelets dorsally compressed; upper lemma without a crest:
78. Spikelets with or without straight or flexuous hairs .. .. **63. Panicum**
78. Spikelets armed with hooked hairs . **64. Ancistrachne**
74. Spikelets subtended by 1 or more bristles, spines or scales:
79. Bristles persistent on the branches; upper lemma indurated, glabrous (if coriaceous with pubescent margins see *Pennisetum glaucum*) .... **75. Setaria**
79. Bristles or scales forming an involucre, deciduous with the spikelets:
80. Involucre composed of bristles:
81. Bristles free throughout, more or less filiform .... **82. Pennisetum**
81. Bristles flattened and connate below to form a cup (disc in *C. ciliaris*), often spinous .... **83. Cenchrus**
80. Involucre composed of stiff glumaceous scales .... **84. Anthephora**
73. Inflorescence of more or less unilateral racemes:
82. Racemes short (1–8 spikelets with filiform rhachis extension) more or less embedded in a thickened axis .... **79. Stenotaphrum**
82. Racemes free from the axis, appressed or divergent:
83. Spikelets bisexual below, male above; rhachis of the single raceme foliaceous and folded transversely to form a capsule .... **72. Thuarea**
83. Spikelets all bisexual:
84. Upper lemma cartilaginous with thin flat margins covering most of the palea, awnless .... **81. Digitaria**
84. Upper lemma coriaceous to crustaceous (chartaceous and awned in *Alloteropsis*) with narrow inrolled margins clasping only the edges of the palea:
85. Lower glume with a long awn; spikelets laterally compressed .... **62. Oplismenus**
85. Lower glume at most with a brief awn-point; spikelets dorsally compressed:
86. Spikelets with a bead-like swelling at the base .... **70. Eriochloa**
86. Spikelets passing smoothly into the pedicel:
87. Lower glume well-developed in all spikelets:
88. Upper lemma pubescent ...... **71. Entolasia**
88. Upper lemma glabrous:

89. Raceme rhachis terminating in a bare point or bristle:
    90. Spikelets not subtended by a bristle; palea of lower floret, when present, permanently hyaline .... **76. Paspalidium**
    90. Spikelets each subtended by a viscid bristle; palea of lower floret developing coriaceous flanks and winged keels at maturity .... **77. Ixophorus**
89. Raceme rhachis terminating in a spikelet:
    91. Inflorescence of untidy digitate racemes; upper lemma crisply chartaceous with an awn 1.5–3 mm long .... **68. Alloteropsis**
    91. Inflorescence of racemes borne along a central axis; upper lemma indurate, awnless or sometimes with a mucro up to 1.2 mm long in *Urochloa*:
        92. Racemes usually 4-rowed; spikelets gibbously plano-convex, cuspidate to awned; tip of palea reflexed and slightly protuberant .... **67. Echinochloa**
        92. Racemes 1–2-rowed; spikelets plump, obtuse to acute; tip of palea not reflexed .... **69. Urochloa**
87. Lower glume absent or minute (occasionally developed in some spikelets of *Paspalum macrophyllum*):
    93. Spikelets paired, or single but then back of upper lemma facing rhachis; upper lemma strongly plano-convex and often orbicular .... **73. Paspalum**
    93. Spikelets single, back of upper lemma turned away from rhachis; upper lemma elliptic or oblong .... **74. Axonopus**
71. Glumes thicker than upper lemma, this hyaline to membranous; lower glume as long as spikelet or almost so:
    94. Spikelets bisexual, though often paired with male or barren spikelets:
        95. Upper lemma awned from low down on the back; spikelets laterally compressed, accompanied by a minute barren pedicel .... **107. Arthraxon**
        95. Upper lemma with or without an awn from tip or sinus of bilobed tip:

96. Spikelets all alike or differing slightly in size, unaccompanied by barren pedicels:
  97. Racemes single:
    98. Spikelets in threes, 2 sessile, 1 pedicelled; upper glume awnless . . . . **93. Polytrias**
    98. Spikelets paired; upper glume awned . . . **94. Pogonatherum**
  97. Racemes two to many:
    99. Callus conspicuously bearded:
      100. Rhachis of raceme fragile; one spikelet in the pair sessile, the other pedicelled . . . . **89. Saccharum**
      100. Rhachis of raceme not disarticulating; both spikelets pedicelled:
        101. Racemes digitate, or (in our species) on an axis not much longer than the racemes, these 10–25 cm long . . . . **90. Miscanthus**
        101. Racemes borne along a central axis in a contracted or spiciform inflorescence, the branches 0.5–6 cm long . . . . **91. Imperata**
    99. Callus glabrous to puberulous:
      102. Spikelets single, laterally compressed . . . **103. Dimeria**
      102. Spikelets paired, dorsally compressed:
        103. Lower glume of sessile spikelet convex, ungrooved, villous . . . . **92. Eulalia**
        103. Lower glume of sessile spikelet with a median groove, glabrous on the back . . . . **95. Microstegium**

96. Spikelets different, the sessile fertile, the pedicelled male, barren or vestigial (sometimes fertile in *Ischaemum*, but then differing in shape from the sessile):
  104. Pedicels and internodes with a translucent median line:
    105. Racemes of few (1–8) spikelets, in a loose branched panicle . . . . **99. Capillipedium**
    105. Racemes of many spikelets, digitate or borne along a central axis . . . . **100. Bothriochloa**
  104. Pedicels and internodes solid:
    106. Inflorescence a panicle with elongated central axis:
      107. Lower glume of sessile spikelets dorsally compressed . . . . **96. Sorghum**
      107. Lower glume of sessile spikelets laterally compressed . . . . **97. Chrysopogon**
    106. Inflorescence of single, paired or digitate racemes:
      108. Racemes paired or digitate:
        109. Upper lemma awned from entire tip . . . . **98. Dichanthium**
        109. Upper lemma awned from sinus of bidentate tip or awnless:

110. Lower floret of sessile spikelet male with palea; pedicels and internodes typically displaying a U-shape on back of raceme .... **101. Ischaemum**

110. Lower floret of sessile spikelet barren without a palea:

111. Lower glume of sessile spikelet 2-keeled:

112. Racemes exserted, not deflexed at maturity; leaves not aromatic .. **104. Andropogon**

112. Racemes more or less enclosed in spatheoles, deflexed at maturity; leaves aromatic when chewed .... **105. Cymbopogon**

111. Lower glume of sessile spikelet rounded on back .... **108. Hyparrhenia**

108. Racemes single:

113. Racemes with 1(– 3 in *Themeda intermedia*) fertile spikelet:

114. Fertile spikelet accompanied only by 2 pedicelled spikelets ..... **102. Apluda**

114. Fertile spikelet surrounded by an involucre of 4 basal sterile spikelets, and accompanied by pedicelled spikelets .. .. **110. Themeda**

113. Racemes with many fertile spikelets:

115. Upper lemma of sessile spikelet awned (rarely awnless but then rhachis internode tip cupuliform):

116. Callus of sessile spikelet ferociously pungent ..... **109. Heteropogon**

116. Callus of sessile spikelet obtuse:

117. Upper lemma entire, awned ... . **98. Dichanthium**

117. Upper lemma bilobed, rarely entire and awnless .... **106. Schizachyrium**

115. Upper lemma of sessile spikelet awnless:

118. Pedicels free:

119. Pedicelled spikelet well-developed .. **111. Elionurus**

119. Pedicelled spikelet absent, but represented by a foliaceous pedicel . . . **112. Eremochloa**

118. Pedicels fused to internode; raceme smoothly cylindrical; callus truncate with a central peg:

120. Raceme rhachis tough; pedicelled spikelet not much different from sessile . . . . **116. Hemarthria**

120. Raceme rhachis fragile:

121. Sessile spikelet globose, its lower glume rugose or latticed . . . . **114. Hackelochloa**

121. Sessile spikelet oblong or ovate, its lower glume smooth:

122. Pedicelled spikelet developed, 1–6 mm long; sessile spikelets single . . . . . **113. Rottboellia**

122. Pedicelled spikelet absent or vestigial; sessile spikelets single, or in opposite pairs with a pedicel between them . . . . **115. Mnesithea**

94. Spikelets unisexual, in separate inflorescences or separate parts of the same inflorescence:

123. Inflorescence (the female in *Zea*) subtended by a herbaceous sheath:

124. Racemes bisexual, female below, male above:

125. Female spikelet partially sunk in the broad rhachis . . . . **117. Tripsacum**

125. Female spieklet embracing the filiform rhachis . . . . **119. Chionachne**

124. Racemes unisexual, the lateral forming female "cobs", the terminal a digitate male "tassel" . . . . **118. Zea**

123. Inflorescence enclosed in a bony ovoid utricle . . . . **120. Coix**

Enumeration of species

PHAREAE

1. Scrotochloa

S. urceolata (Roxb.) Judz., *Phytologia* 56: 300 (1984). *Pharus urceolatus* Roxb., *Fl. Ind. ed. 1832*, 3: 611 (1832). *Leptaspis urceolata* (Roxb.) R. Br. in Benn., *Pl. Jav. Rar.*: 23 (1838); *Fl. Sol.* 185.
SOL; tropical Asia, Australia.

2. Leptaspis

1. Leaf-blades linear, 20–45 cm long, 4–10 mm wide; female spikelets 2.5–4 mm long .... **L. angustifolia**
1. Leaf-blades lanceolate to ovate, 10–25 cm long, 10–60 mm wide; female spikelets 5–7 mm long:
   2. Inflorescence branches solitary at each node, erect, 1–5 cm long; leaf-blades lanceolate .... **L. banksii**
   2. Inflorescence branches mostly 2–3 at the lower nodes, ascending, 4–12 cm long; leaf-blades oblong to ovate .... **L. zeylanica**

L. angustifolia Summerh. & C. E. Hubb., *Bull. Misc. Inform., Kew* 1927: 40, 78 (1927); *Fl. Fiji* 322.
FIJ; New Guinea.

L. banksii R. Br., *Prodr.*: 211 (1810); *Fl. Sol.*: 185. *L. lanceolata* Zoll., *Syst. Verz.*: 53 (1854), nom. nud.; *Fl. N. Cal.* 27.
NWC, SOL; Malesia, New Guinea, Australia.

L. zeylanica Steud., *Syn. Pl. Glumac.* 1: 8 (1853). *L. cochleata* Thwaites, *Enum. Pl. Zeyl.*: 357 (1864), nom. superfl.; *Fl. Sol.* 185.
SOL; Africa, tropical Asia.

ORYZEAE

3. Oryza

1. Spikelets persistent, hispidulous .... **O. sativa**
1. Spikelets deciduous, bearing hooked spines .... **O. neocaledonica**

O. neocaledonica Morat, *Bull. Mus. Natl. Hist. Nat, ser. 4, Sect. B, Adansonia* 16: 3 (1994).
NWC.

O. sativa L., *Sp. Pl.* 1: 333 (1753).
Rice: cultivated throughout tropical and warm temperate regions.

4. Zizania

Z. latifolia (Griseb.) Stapf, *Bull. Misc. Inform., Kew* 1909: 385 (1909); *Hydropyrum latifolium* Griseb. in Ledeb., *Fl. Ross.* 4: 466 (1852).
HAW; eastern Asia.

ERHARTEAE

5. Ehrharta

1. Sterile lemmas obtuse to acute, awnless, the upper auriculate at the base:
   2. Sterile lemmas with long hairs on sides, keel or margin, smooth .... **E. calycina**
   2. Sterile lemmas glabrous or scabrous, often rugose .... **E. erecta**
1. Sterile lemmas attenuate, awned, not auriculate:
   3. Fertile lemma bidentate, awnless; callus of upper sterile lemma puberulous .... **E. diplax**
   3. Fertile lemma acuminate or shortly awned; callus of upper sterile lemma glabrous; glumes reduced to obscure lips .... **E. stipoides**

E. calycina *Sm., Pl. Icon. Ined.* t. 33 (1789).
HAW; South Africa.

E. diplax F. Muell., *Fragm.* 7: 90 (1870); *Fl. Soc.* 336. *Diplax avenacea* Raoul, *Ann. Sci. Nat., Bot.*, sér. 3, 2: 116 (1844); *Microlaena avenacea* (Raoul) Hook. f., *Handb. N. Zeal. Fl.*: 320 (1864); *Fl. Fiji* 320.
FIJ, SCI; New Guinea, New Zealand.

E. erecta Lam., *Encycl.* 2(1): 347 (1786).
HAW; Africa.

E. stipoides Labill., *Nov. Holl. Pl.* 1: 91 (1804); *Man. Haw.* 1536; *Easter* 77. *Microlaena stipoides* (Labill.) R. Br., *Prodr.*: 210 (1810).
EAS, HAW; Southeast Asia, Australia, New Zealand.

Stipeae

6. Stipa

1. Lemma with a ring of hairs 0.5–1 mm long surrounding base of awn; spikelets 12–19 mm long . . . . **S. cernua**
1. Lemma without a distinct ring of hairs around base of awn; spikelets 8–10 mm long . . . . **S. scabra**

S. cernua Stebbins & Löve, *Madroño* 6: 137 (1941); *Man. Haw.* 1599. *Nassella cernua* (Stebbins & Löve) Barkworth, *Taxon* 39: 609 (1990).
HAW; California, Mexico.

S. scabra Lindl. in T. Mitch., *J. Exped. Trop. Australia*: 31 (1848); *Easter* 83. *Stipa horridula* Pilg. in Skottsb., *Nat. Hist. Juan Fernandez* 2: 64 (1922).

7. Oryzopsis

O. miliacea (L.) Asch. & Schweinf., *Mém. Inst. Égypt.* 2: 169 (1887). *Agrostis miliacea* L., *Sp. Pl.* 1: 61 (1753). *Piptatherum miliaceum* (L.) Coss., *Notes Pl. Crit.*: 129 (1812).
HAW; Mediterranean, Middle East and widely introduced.

Poeae

8. Festuca

1. Leaf-blades setaceous, 0.5–2 mm wide; lemma with an awn 1–5 mm long . . . . **F. rubra**
1. Leaf-blades flat, 3–12 mm wide; lemma awnless or with an awn up to 4 mm long:
   2. Sheath without auricles; rhachilla puberulous . . . . **F. hawaiiensis**
   2. Sheath with falcate auricles; rhachilla glabrous; panicle branches in pairs:
      3. Auricles glabrous; shorter branch of the pair bearing 1–2 spikelets . . . **F. pratensis**
      3. Auricles ciliolate; shorter branch of the pair bearing 3 or more spikelets . . . . **F. arundinacea**

F. arundinacea Schreb., *Spic. Fl. Lips.*: 57 (1771), nom. conserv.; *Man. Haw.* 1547. *Schedonorus arundinaceus* (Schreb.) Dumort., *Observ. Gramin. Belg.*: 106 (1823), non Roem. & Schult. (1817).
HAW; temperate Old World and widely introduced.

F. hawaiiensis Hitchc., *Mem. Bernice Pauahi Bishop Mus.* 8: 115, t. 4 (1922); *Man. Haw.* 1547.
HAW.

F. pratensis Huds., *Fl. Angl.*: 37 (1762). *Schedonorus pratensis* (Huds.) P. Beauv., *Ess. Agrostogr.*: 99 (1812).
HAW; temperate Old World.

F. rubra L., *Sp. Pl.* 1: 74 (1753); *Man. Haw.* 1547. *F. briquetii* St.-Yves in *Rev. Bretonne Bot. Pure Appl.* 2: 16, 53 (1927).
HAW; temperate Old World, North America.

9. Vulpia

1. Lower glume $^{1}/_{2}$–$^{3}/_{4}$ length of upper; upper glume 3-veined; panicle lanceolate to narrowly oblong, exserted .... **V. bromoides**
1. Lower glume $^{1}/_{6}$ to nearly $^{1}/_{2}$ length of upper; upper glume 1(– 3)-veined; panicle linear, often embraced by the uppermost sheath .... **V. myuros**

V. bromoides (L.) Gray, *Nat. Arr. Brit. Pl.* 2: 124 (1821); *Man. Haw.* 1603. *Festuca bromoides* L., *Sp. Pl.* 1: 75 (1753).
HAW; Europe and widely introduced.

V. myuros (L.) C. C. Gmel., *Fl. Bad.* 1: 8 (1805); *Man. Haw.* 1603; *Easter* 88. *Festuca myuros* L., *Sp. Pl.* 1: 74 (1753).
EAS, HAW, TUB; Europe, western Asia and widely introduced.

10. Lolium

1. Upper glume reaching or exceeding the uppermost lemma; florets elliptic to ovate, turgid .... **L. temulentum**
1. Upper glume much shorter than rest of the spikelet; florets oblong or lanceolate-oblong, not turgid:
   2. Lemma awned; annual .... **L. multiflorum**
   2. Lemma awnless; perennial .... **L. perenne**

L. multiflorum Lam., *Fl. Franç.* 3: 621 (1778); *Man. Haw.* 1561.
HAW; temperate Old World and widely introduced.

L. perenne L., *Sp. Pl.* 1: 83 (1753); *Man. Haw.* 1561; *Easter* 78.
EAS, HAW; temperate Old World and widely introduced.

L. temulentum L., *Sp. Pl.* 1: 83 (1753); *Man. Haw.* 1560.
HAW; eastern Europe, temperate Asia and widely introduced.

11. Lamarckia

L. aurea (L.) Moench, *Methodus*: 201 (1794). *Cynosurus aureus* L., *Sp. Pl.* 1: 73 (1753).
HAW; Mediterranean, western Asia and widely introduced.

12. Briza

1. Spikelets 14–25 mm long, 8–15 mm wide .... **B. maxima**
1. Spikelets 3–5 mm long, 3–6 mm wide .... **B. minor**

B. maxima L., *Sp. Pl.* 1: 70 (1753); *Man. Haw.* 1504.
HAW, MRQ; Europe and widely introduced.

B. minor L., *Sp. Pl.* 1: 70 (1753); *Fl. Poly.* 85; *Man. Haw.* 1505; *Easter* 73.
EAS, HAW, MRQ, TUB; Europe, western Asia and widely introduced.

13. Poa

1. Plants annual .... **P. annua**
1. Plants perennial:
   2. Sheath margins free or connate up to half their length; ligule as wide as blade, not extending onto shoulders of sheath; rhizomatous:
      3. Plants dioecious; panicle densely contracted; lemmas 4.2–6.5 mm long .... **P. arachnifera**
      3. Plants bisexual:
         4. Culms elliptic in section; panicle open or contracted; lemmas 2.5–3 mm long ... . **P. compressa**
         4. Culms circular in section; panicle open; lemmas 3–4 mm long .... **P. pratensis**
   2. Sheath margins connate to the top; ligule surrounding culm; rhizomes absent:
      5. Ligule deeply fimbriate; panicle contracted, its branches 0.5–2 cm long ... **P. mannii**
      5. Ligule entire or dentate-serrate:
         6. Panicle effuse, the longer branches usually 5–10 cm long; culms erect or ascending .... **P. sandvicensis**
         6. Panicle compact, the longer branches 2–4 cm long; culms often decumbent and cascading down banks .... **P. siphonoglossa**

P. annua L., *Sp. Pl.* 1: 68 (1753); *Fl. N. Cal.* 35; *Man. Haw.* 1583; *Easter* 81.
EAS, HAW, NWC; cosmopolitan in temperate regions and tropical highlands.

P. arachnifera Torr. in Marcy, *Exp. Red. Riv. Louis.* Bot.: 301 (1853); *List Micro.* 56.
MRN; southern USA.

P. compressa L., *Sp. Pl.* 1: 69 (1753).
HAW; temperate regions.

P. mannii Munro ex Hillebr., *Fl. Hawaiian Isl.*: 526 (1888); *Man. Haw.* 1584.
HAW.

P. pratensis L., *Sp. Pl.* 1: 67 (1753); *Man. Haw.* 1584.
HAW; temperate Old World and widely introduced.

P. sandvicensis (Reichardt) Hitchc., *Mem. Bernice Pauahi Bishop Mus.* 8: 121 (1922); *Man. Haw.* 1534. *Festuca sandvicensis* Reichardt, *Sitzungber. Kaiserl. Akad. Wiss., Math.-Naturwiss. Cl.* 76(1): 726 (1878). *Poa longeradiata* Hillebr., *Fl. Hawaiian Isl.*: 526 (1888).
HAW.

P. siphonoglossa Hack., *Repert. Spec. Nov. Regni Veg.* 11: 24 (1912); *Man. Haw.* 1585.
HAW.

14. Dactylis

D. glomerata L., *Sp. Pl.* 1: 71 (1753); *Fl. Fiji* 298; *Man. Haw.* 1521.
FIJ, HAW, MRQ; Europe and widely introduced.

MELICEAE

15. Glyceria

G. notata Chevall., *Fl. Gén. Env. Paris* 2: 174 (1827). *G. fluitans* subsp. *plicata* Fr., Novit. Fl. Suec. Mant. 2: 6 (1839). *G. plicata* (Fr.) Fr., *Novit. Fl. Suec. Mant.* 3: 176 (1845). *G. fluitans sensu* Herbst & Wagner, *Occas. Pap. Bernice Pauahi Bishop Mus.* 58: 27 (1999).
HAW; Europe, Middle East.

AVENEAE

16. Arrhenatherum

A. elatius (L.) P. Beauv. ex J. & C. Presl, *Fl. Čech.*: 17 (1819). *Avena elatior* L., *Sp. Pl.* 1: 79 (1753).
HAW; temperate Old World and widely introduced.

17. Avena

1. Rhachilla and lemma glabrous; florets persistent .... **A. sativa**
1. Rhachilla and lemma hairy; florets deciduous:
   2. Lemma tip dentate, without bristles .... **A. fatua**
   2. Lemma tipped by 2 bristles 3–12 mm long .... **A. barbata**

A. barbata Pott ex Link, *J. Bot.* (*Schrader*) 2: 315 (1799).
HAW; southern Europe, western Asia and widely introduced.

A. fatua L., *Sp. Pl.* 1: 80 (1753); *Man. Haw.* 1499; *Easter* 73.
EAS, HAW, MRQ, SCI; temperate Old World and widely introduced.

A. sativa L., *Sp. Pl.* 1: 79 (1753); *Fl. N. Cal.* 33; *Fl. Guam* 196; *Fl. Fiji* 315; *List Micro.* 31; *Man. Haw.* 1500.
FIJ, HAW, MRN, MRQ, MRS, NWC; cultivated oats of temperate regions.

18. Koeleria

K. macrantha (Ledeb.) Schult., *Mant.* 2: 345 (1824). *Aira macrantha* Ledeb., *Mém. Acad. Imp. Sci. St. Pétersbourg*, Sér. 7, 5: 515 (1812). *Koeleria nitida* Nutt., *Gen. N. Amer. Pl.* 1: 74 (1818); *Man. Haw.* 1557.
HAW; temperate Old World and widely introduced.

19. Trisetum

1. Glumes subequal, 5–6 mm long; awn arising $^{3}/_{4}$ way up back of lemma, recurved .... **T. glomeratum**
1. Glumes unequal, the lower 3–4 mm long:
   2. Awn straight or weakly curved, arising just below apex of lemma ..... **T. inaequale**
   2. Awn geniculate, arising $^{1}/_{2}$–$^{2}/_{3}$ way up back of lemma .... **T. flavescens**

T. flavescens (L.) P. Beauv., *Ess. Agrostogr.*: 88 (1812). *Avena flavescens* L., *Sp. Pl.* 1: 80 (1753).
HAW; temperate regions.

T. glomeratum (Kunth) Trin. ex Steud., *Nomencl. Bot.*, ed. 2 2: 714 (1841); *Man. Haw.* 1602. *Koeleria glomerata* Kunth, *Enum. Pl.* 1: 526 (1833). *K. vestita* Steud., *Syn. Pl. Glumac.* 1: 294 (1854).
HAW.

T. inaequale Whitney, *Occas. Pap. Bernice Pauahi Bishop Mus.* 13: 171 (1937); *Man. Haw.* 1602.
HAW.

20. Deschampsia

1. Awn arising $^{1}/_{4}$–$^{1}/_{2}$ way up back of lemma; floret callus hairs 1.5–2.5 mm long .. **D. klossii**
1. Awn arising near base of lemma; floret callus hairs 1 mm long .... **D. nubigena**

D. klossii Ridl., *Bull. Misc. Inform.*, Kew 1913: 268 (1913).
SCI; New Guinea.

D. nubigena Hillebr., *Fl. Hawaiian Isl.*: 521 (1888). *Man. Haw.* 1524; *Fl. Soc.* 333. *Aira australis* Steud., *Syn. Pl. Glumac.* 1: 220 (1854), non Raoul (1846). *Deschampsia australis* Hillebr., *Fl. Hawaiian Isl.*: 520 (1888). *D. pallens* Hillebr., l.c.: 520. *Aira nubigena* (Hillebr.) Hitchc., *Mem. Bernice Pauahi Bishop Mus.* 8: 145 (1922). *A. hawaiiensis* Skottsb., *Acta Horti Gothob.* 2: 205 (1926). *A. hawaiiensis* f. *depauperata* Skottsb., l.c.: 207. *A. hawaiiensis* f. *haleakalensis* Skottsb., l.c.: 207. *A. pallida* Skottsb., l.c.: 204. *A. pallida* var. *tenuissima* Skottsb., l.c.: 203. *Deschampsia australis* f. *haleakalensis* (Skottsb.) Skottsb., *Acta Horti Gothob.* 15: 279 (1944). *D. australis* subsp. *nubigena* (Hillebr.) Skottsb., l.c.: 279. *D. australis* var. *gracilis* Skottsb., l.c.: 280. *D. australis* var. *tenuissima* (Skottsb.) Skottsb., l.c.: 280. *D. hawaiiensis* (Skottsb.) H. St.John, *Bull. Torrey Bot. Club* 72: 23 (1944), nom. superfl.
CRL, HAW, SCI.

21. Aira

A. caryophyllea L., *Sp. Pl.* 1: 66 (1753); *Man. Haw.* 1496.
HAW, MRQ; Europe and widely introduced.

22. Holcus

H. lanatus L., *Sp. Pl.* 2: 1048 (1753); *Man. Haw.* 1551.
HAW; Europe, western Asia and widely introduced.

23. Anthoxanthum

A. odoratum L., *Sp. Pl.* 1: 28 (1753); *Man. Haw.* 1498.
HAW, MRQ; temperate Old World and widely introduced.

24. Phalaris

1. Spikelets in clusters of 6–7, only the central one fertile, the cluster falling as a whole ... . **P. paradoxa**
1. Spikelets single, all fertile, their glumes persisting after the florets have fallen:
   2. Sterile lemmas 2, equal, broad and chaffy, $^1/_2$–$^2/_3$ length of fertile; annual .... **P. canariensis**
   2. Sterile lemma 1, sometimes 2 but then markedly unequal, subulate or scale-like, less than $^1/_2$ length of fertile:
      3. Glume wings toothed or erose; sterile lemma 1, often reduced to a tiny scale; annual .... **P. minor**
      3. Glume wings entire; sterile lemmas 1 or 2, well developed; perennial ... **P. aquatica**

P. aquatica L., *Cent. Pl.* I: 4 (1755). *P. tuberosa* L., *Mant. Pl.* 2: 557 (1771); *Fl. Guam* 198. *P. stenoptera* Hack., *Repert. Spec. Nov. Regni Veg.* 5: 333 (1908). *P. tuberosa* var. *stenoptera* (Hack.) Hitchc., *J. Wash. Acad. Sci.* 24: 292 (1934); *List Micro.* 55.
HAW, MRN; Mediterranean and widely introduced.

P. canariensis L., *Sp. Pl.* 1: 54 (1753).
HAW; Europe and widely introduced.

P. minor Retz., *Observ. Bot.* 3: 8 (1783); *Fl. N. Cal.* 33.
HAW, NWC; temperate Old World and widely introduced.

P. paradoxa L., *Sp. Pl.*, ed. 2, 2: 1665 (1763).
HAW; temperate Old World and widely introduced.

25. Agrostis

1. Lemma pilose; rhachilla extended beyond the floret as a ciliate bristle .... **A. avenacea**
1. Lemma glabrous; rhachilla not extended beyond the floret:

2. Palea $^1/_2$–$^2/_3$ length of lemma; lemma usually awnless:
   3. Ligules of vegetative shoots as long as or longer than wide; panicle contracted after flowering; stoloniferous .... **A. stolonifera**
   3. Ligules of vegetative shoots shorter than wide; panicle loose; rhizomatous .... **A. capillaris**
2. Palea less than $^1/_4$ length of lemma; lemma awned or awnless:
   4. Panicle branches, at least the lower, naked towards the base; panicle open or contracted, lanceolate to ovate; culms ascending or decumbent; stoloniferous; awn, when present, arising near base of lemma .... **A. canina**
   4. Panicle branches bearing spikelets to the base; panicle contracted to spiciform; culms erect, tufted, without stolons; awn, when present, arising $^1/_2$ way up lemma:
      5. Leaf-blades flat, 2–8 mm wide; panicle contracted, linear to lanceolate .... **A. exarata**
      5. Leaf-blades involute, 1–2 mm wide; panicle spiciform, linear . . . **A. sandwicensis**

A. avenacea Gmel., *Syst. Nat.* 1: 171 (1791); *Man. Haw.* 1492; *Easter* 71. *Avena filiformis* G. Forst., *Fl. Ins. Austr.*: 9 (1786), non *Agrostis filiformis* Vill. (1787). *Lachnagrostis filiformis* (G. Forst.) Trin., *Fund. Agrost.*: 128 (1820). *L. chamissonis* Trin., *Gram. Unifl. Sesquifl.*: 216 (1824).
EAS, HAW; New Guinea, Australia, New Zealand.

A. canina L., *Sp. Pl.* 1: 62 (1753).
HAW; Europe and widely introduced.

A. capillaris L., *Sp. Pl.* 1: 62 (1753).
HAW, NWC; temperate Old world and widely introduced.

A. exarata Trin., *Gram. Unifl. Sesquifl.*: 205 (1824).
HAW, SAM; North America.

A. sandwicensis Hillebr., *Fl. Hawaiian Isl.*: 515 (1888); *Man. Haw.* 1494. *A. fallax* Hillebr., l.c.: 516. *A. rockii* Hack., *Repert. Spec. Nov. Regni Veg.* 10: 167 (1911).
HAW.

A. stolonifera L., *Sp. Pl.* 1: 62 (1753); *Man. Haw.* 1494; *Easter* 72.
EAS, HAW, MRQ; temperate Old World and widely introduced

26. Calamagrostis

1. Callus hairs 4–5 mm long; culms robust, 50–200 cm long .... **C. expansa**
1. Callus hairs 0.5 mm long; culms 30–50 cm long .... **C. hillebrandii**

C. expansa (Munro ex Hillebr.) Hitchc., *Mem. Bernice Pauahi Bishop Mus.* 8: 149 (1922); *Man. Haw.* 1509. *Deyeuxia expansa* Munro ex Hillebr., *Fl. Hawaiian Isl.*: 519 (1888).
HAW.

C. hillebrandii (Munro ex Hillebr.) Hitchc., *Mem. Bernice Pauahi Bishop Mus.* 8: 147 (1922); *Man. Haw.* 1509. *Deyeuxia hillebrandii* Munro ex Hillebr., *Fl. Hawaiian Isl.*: 519 (1888).
HAW.

27. Ammophila

A. arenaria (L.) Link, *Hort. Berol.* 1: 105 (1827); *Fl. Fiji* 315. *Arundo arenaria* L., *Sp. Pl.* 1: 82 (1753).
FIJ, HAW; Europe, western Asia.

28. Dichelachne

1. Awn 20–40 mm long, inserted 1–3 mm below lemma tip; lemma 3.75–8 mm long ..... **D. crinita**
1. Awn 8–18 mm long, inserted 0.5–1 mm below lemma tip; lemma 2.75–4 mm long ..... **D. micrantha**

D. crinita (L. f.) Hook. f., *Fl. Nov.-Zel.* 1: 293 (1853); *Easter* 75. *Anthoxanthum crinitum* L.f., *Suppl. Pl.*: 90 (1781). *Agrostis rapensis* F. Br., *Bernice P. Bishop Mus. Bull.* 84: 80 (1931).
EAS, HAW, NUE, TUB; New Guinea, Australia, New Zealand.

D. micrantha (Cav.) Domin, *Biblioth. Bot.* 85: 353 (1915); *Easter* 76. *Stipa micrantha* Cav., *Icon.* 5: 42 (1799).
EAS, HAW, TUB; Australia, New Zealand.

29. Gastridium

G. ventricosum (Gouan) Schinz & Thell., *Vierteljahrsschr. Naturf. Ges. Zürich* 58: 39 (1913); *Man. Haw.* 1550; *Easter* 78. *Agrostis ventricosa* Gouan, *Hortus Monsp.* 39, 547 (1762).
EAS, HAW; southern Europe, Southwest Asia.

30. Polypogon

1. Glumes awnless; pedicel wholly deciduous ..... **P. viridis**
1. Glumes awned:
    2. Plant perennial; glume awns 1.5–3 mm long; pedicel wholly deciduous ..... **P. interruptus**
    2. Plants annual:
        3. Glume awns 4–7 mm long, at least twice length of glume; pedicel disarticulating towards the top, leaving a persistent stump ..... **P. monspeliensis**
        3. Glume awns 0.6–3 mm long, half to one and a half times length of glume; pedicel wholly deciduous ..... **P. fugax**

P. fugax Steud., *Syn. Pl. Glumac.* 1: 184 (1854).
HAW; temperate Asia.

P. interruptus Kunth in Humb. & Bonpl., *Nov. Gen. Sp.* 1: 134 (1815); *Man. Haw.* 1585.
HAW; South America.

P. monspeliensis (L.) Desf., *Fl. Atlant.* 1: 66 (1798); *Man. Haw.* 1586. *Alopecurus monspeliensis* L., *Sp. Pl.* 1: 61 (1753).
HAW; southern Europe, temperate Asia and widely introduced.

P. viridis (Gouan) Breistr., *Bull. Soc. Bot. France* 110 (Sess. Extr.): 56 (1966). *Agrostis viridis* Gouan, *Hortus Monsp.*: 546 (1762). *Phalaris semiverticillata* Forssk., *Fl. Aegypt.-Arab.*: 17 (1775). *Agrostis verticillata* Vill., *Prosp. Hist. Pl. Dauphiné*: 16 (1779); *Fl. N. Cal.* 33. *A. semiverticillata* (Forssk.) C. Chr., *Dansk. Bot. Ark.* 4(3): 12 (1922); *Man. Haw.* 1494.
HAW, NWC; southern Europe, temperate Asia and widely introduced.

31. Alopecurus

A. utriculatus Banks & Sol., *Russ. Nat. Hist. Aleppo*, ed. 2, 2: 243 (1794). *A. anthoxanthoides* Boiss., *Diagn. Pl. Orient.* 2(13): 42 (1854). *Fl. N. Cal.* 33.
NWC; western Asia.

32. Phleum

P. pratense L., *Sp. Pl.* 1: 59 (1753).
HAW; temperate Old World and widely introduced.

Bromeae

33. Bromus

1. Lemmas strongly laterally compressed and keeled .... **B. catharticus**
1. Lemmas rounded on the back:
   2. Lower glume 3–7-veined; spikelets lanceolate to ovate, tapering towards the top; lemmas awned:
      3. Lemma margins inrolled below the middle, the lemma 7–9 mm long; leaf-sheaths usually glabrous .... **B. secalinus**
      3. Lemma margins flat, the lemma 8–11 mm long; leaf-sheaths hairy ... **B. hordeaceus**
   2. Lower glume 1-veined; spikelets oblong or cuneate, gaping at the top:
      4. Lemmas 20–35 mm long .... **B. diandrus**
      4. Lemmas 9–20 mm long:
         5. Panicle drooping, open, the branches mostly longer than spikelets:
            6. Panicle branches simple, each bearing 1(– 3) spikelets .... **B. sterilis**
            6. Panicle branches divided, each bearing at least 4 spikelets ..... **B. tectorum**
         5. Panicle erect, dense, the branches mostly shorter than spikelets:
            7. Panicle loose; branches partly visible, 10–30 mm long ..... **B. madritensis**
            7. Panicle densely contracted; branches hidden, 1–10 mm long ..... **B. rubens**

B. catharticus Vahl, *Symb. Bot.* 2: 22 (1791); *Easter* 74. *B. unioloides* (Willd.) Raspail, *Ann. Sci. Nat. (Paris)* 5: 439 (1825) non Kunth (1815); *Fl. N. Cal.* 35. *B. willdenowii* Kunth, *Révis. Gramin.* 1: 134 (1829); *Man. Haw.* 1508.
EAS, HAW, NWC; South America and widely introduced.

B. diandrus Roth, *Bot. Abh. Beobacht.*: 44 (1787). *B. rigidus* Roth, *Bot. Mag (Römer & Usteri)* 4(10): 21 (1790); *Man. Haw.* 1508. *B. diandrus* var. *rigidus* (Roth) Sales, *Edinburgh J. Bot.* 50: 9 (1993).
HAW; Europe, western Asia and widely introduced.

B. hordeaceus L., *Sp. Pl.* 1: 77 (1753). *B. mollis* L., *Sp. Pl.*, ed. 2, 1: 112 (1762); *Man. Haw.* 1507.
HAW; Europe and widely introduced.

B. madritensis L., *Cent. Pl.* I: 5 (1755); *Man. Haw.* 1505.
HAW; Europe, western Asia and widely introduced.

B. rubens L., *Cent. Pl.* I: 5 (1755); *Man. Haw.* 1507.
HAW; Europe, western Asia and widely introduced.

B. secalinus L., *Sp. Pl.* 1: 76 (1753).
HAW; Europe and widely introduced.

B. sterilis L., *Sp. Pl.* 1: 77 (1753).
HAW; Europe, western Asia and widely introduced.

B. tectorum L., *Sp. Pl.* 1: 77 (1753).
HAW; Europe, western Asia and widely introduced.

TRITICEAE

34. Elymus

E. repens (L.) Gould, *Madroño* 9: 127 (1947). *Triticum repens* L., *Sp. Pl.* 1: 86 (1753). *Agropyron repens* (L.) P. Beauv., *Ess. Agrostogr.*: 102 (1812); *List Micro.* 30.
CRL; widespread in temperate regions.

35. Triticum

T. aestivum L., *Sp. Pl.* 1: 85 (1753); *Fl. Fiji* 298. *T. sativum* Lam., *Fl. Franç.* 3: 625 (1778); *Fl. N. Cal.* 32.
FIJ, HAW, NWC; cultivated wheat of temperate regions.

36. Hordeum

1. Rhachis tough, the spikelets persistent .... **H. vulgare**
1. Rhachis fragile, disarticulating with the spikelets at maturity:
   2. Lemma awn 18–50 mm long; weedy annual .... **H. murinum**
   2. Lemma awn 5–7 mm long; tufted perennial .... **H. brachyantherum**

H. brachyantherum Nevski, *Trudy Bot. Inst. Akad. Nauk S.S.S.R., Ser. 1, Fl. Sist Vyssh. Rast.* 2: 61 (1936).
HAW; North America.

H. murinum L., *Sp. Pl.* 1: 85 (1753); *Easter* 78. *H. leporinum* Link, *Linnaea* 9: 133 (1834); *Man. Haw.* 1552. *H. murinum* subsp. *leporinum* (Link) Arcang., *Comp. Fl. Ital.*: 805 (1882).
EAS, HAW; Europe, western Asia and widely introduced.

H. vulgare L., *Sp. Pl.* 1: 94 (1753); *Fl. N. Cal.* 32; *Fl. Fiji* 299; *Man. Haw.* 1522.
FIJ, HAW, NWC; cultivated barley of temperate regions.

CENTOTHECEAE

37. Centotheca

C. lappacea (L.) Desv., *Nouv. Bull. Sci. Soc. Philom. Paris* 2: 189 (1810); *Fl. Raro.* 17; *Pl. Tonga* 53; *Fl. Guam* 188; *Fl. Niue* 237; *Pl. Samoa* 110; *Fl. Fiji* 297; *List Micro.* 53; *Fl. Sol.* 184; *Fl. Soc.* 331. *Cenchrus lappaceus* L., *Sp. Pl.*, ed. 2, 2: 1488 (1763). *Centotheca latifolia* Trin., *Fund. Agrost.*: 141 (1820); *Fl. Poly.* 84; *Fl. Sol.* 184.
COO, CRL, FIJ, MRN, MRQ, MRS, NUE, NWC, SAM, SCI, SOL, TON, TUB, VAN, WAL; Old World tropics.

38. Lophatherum

L. gracile Brongn. in Duperrey, *Voy. Monde Phan.*: 50 (1831); *List Micro.* 49.
CRL, MRS, SCI; tropical Asia.

ARUNDINEAE

39. Rytidosperma

1. Lemma 3-awned:
   2. Lemma with 2 marginal and sometimes 2 dorsal tufts of hair, otherwise glabrous and shiny; lateral awns 6–8 mm long .... **R. pilosum**
   2. Lemma with 2 transverse rows of hair, otherwise pubescent; lateral awns 2–3 mm long .... **R. biannulare**
1. Lemma 1-awned, the lobes with or without a mucro up to 1 mm long:

3. Lemma pubescent, and with a conspicuous transverse row of hair tufts 2–4 mm long across the back .... **R. semiannulare**
3. Lemma glabrous and shiny, with marginal hair tufts only .... **R. paschale**

R. biannulare (Zotov) Connor & Edgar, *New Zealand J. Bot.* 17: 324 (1979). *Notodanthonia biannularis* Zotov, *New Zealand J. Bot.* 1: 116 (1963). *Austrodanthonia biannularis* (Zotov) H. P. Linder, *Telopea* 7: 270 (1997).
HAW; New Zealand.

R. paschale (Pilg.) C. M. Baeza, *Gayana, Bot.* 47: 84 (1991). *Danthonia paschalis* Pilg. in Skottsb., *Nat. Hist Juan Fernandez* 2: 67 (1922); *Easter* 75.
EAS.

R. pilosum (R. Br.) Connor & Edgar, *New Zealand J. Bot.* 17: 326 (1979). *Danthonia pilosa* R. Br., *Prodr.*: 177 (1810); *Man. Haw.* 1522.
HAW; Australia, New Zealand.

R. semiannulare (Labill.) Connor & Edgar, *New Zealand J. Bot.* 17: 332 (1979). *Arundo semiannularis* Labill., *Nov. Holl. Pl.* 1: 26 (1804). *Danthonia semiannularis* (Labill.) R. Br., *Prodr.*: 177 (1810). *Notodanthonia semiannularis* (Labill.) Zotov, *New Zealand J. Bot.* 1: 116 (1963).
HAW; Australia.

40. Cortaderia

1. Lemma tip continuing into an awn; plants dimorphic; female lemmas plumose, the bisexual sparsely hairy .... **C. selloana**
1. Lemma tip minutely bidentate and mucronate; plants all female (so far as known); lemmas plumose .... **C. jubata**

C. jubata (Lemoine) Stapf, *Bot. Mag.* t. 7607 (1898). *Gynerium jubatum* Lemoine, *Rev. Hort.* 1878: 449 (1878).
HAW; Australia, South America.

C. selloana (Schult. & Schult. f.) Asch. & Graebn., *Syn. Mitteleur. Fl.* 2: 325 (1900). *Arundo selloana* Schult. & Schult.f., Mant. 3: 605 (1827).
HAW; South America and widely grown as an ornamental.

41. Arundo

A. donax L., *Sp. Pl.* 1: 81 (1753); *Fl. N. Cal.* 34; *Pl. Tonga* 53; *Fl. Guam* 188; *Pl. Samoa* 33; *Fl. Fiji* 299; *List Micro.* 31; *Man. Haw.* 1498; *Easter* 72; *Fl. Soc.* 329. *A. versicolor* Mill., *Gard. Dict.*, ed. 8, no. 3 (1768). *A. donax* var. *versicolor* (Mill.) Stokes, *Bot. Mat. Med.* 1: 100 (1812); *Pl. Samoa* 33; *Fl. Fiji* 300; *List Micro.* 31.
COO, CRL, EAS, FIJ, HAW, MRN, MRQ, MRS, NRU, NUE, NWC, SAM, SCI, TON, VAN; Mediterranean, temperate Asia and widely introduced.

42. Phragmites

1. Leaf-blades smooth beneath, the tips filiform and flexuous; rhachilla hairs 6–10 mm long .... **P. australis**
1. Leaf-blades scabrid beneath (at least in upper half), the tips attenuate and stiff; rhachilla hairs 4–7 mm long .... **P. vallatorius**

P. australis (Cav.) Trin. ex Steud., *Nomencl. Bot.*, ed. 2, 2: 324 (1841). *Arundo australis* Cav., *Anales Hist. Nat.* 1: 100 (1799). *Phragmites communis* Trin., *Fund. Agrost.*: 134 (1820); *Fl. Raro.* 21; *Fl. N. Cal.* 34.
COO, HAW, NWC; cosmopolitan in temperate and subtropical regions.

P. vallatorius (L.) Veldkamp, *Blumea* 37: 233 (1992). *Arundo vallatoria* L., *Herb. Amb.*: 15 (1754). *A. karka* Retz., *Observ. Bot.* 4: 21 (1786). *Phragmites karka* (Retz.) Trin. ex Steud., *Nomencl. Bot.*, ed. 2, 2: 324 (1841); *Fl. Guam* 189; *List Micro.* 56; *Fl. Sol.* 185.
COO, CRL, HAW, MRN, MRS, NWC, SOL, VAN; Nigeria to Ethiopia, tropical Asia.

Note. The differences between the two species are minor and some authors have merged them. If separate, then the presence of *Phragmites australis* in the Pacific is rare.

THYSANOLAENEAE

43. Thysanolaena

T. latifolia (Roxb. ex Hornem.) Honda, *J. Fac. Sci. Univ. Tokyo Sect. 3, Bot.* 3: 312 (1930). *Melica latifolia* Roxb. ex Hornem., *Hort. Bot. Hafn. Suppl.*: 117 (1819).
HAW; tropical Asia and often grown as an ornamental.

ARISTIDEAE

44. Aristida

1. Awns disarticulating at tip of a twisted column .... **A. repens**
1. Awns not disarticulating:
   2. Lemma laterally compressed and keeled; usually annual .... **A. adscensionis**
   2. Lemma terete, rounded on back; perennial:
      3. Awns separated from lemma tip by a twisted column 3 mm long; central awn recurved .... **A. novae-caledoniae**
      3. Awns arising from lemma tip, straight:
         4. Culms pubescent below nodes .... **A. pilosa**
         4. Culms glabrous .... **A. ramosa**

A. adscensionis L., *Sp. Pl.* 1: 82 (1753).
HAW; throughout the tropics and subtropics.

A. novae-caledoniae Henrard, *Meded. Rijks-Herb.* 54A: 383 (1927); *Fl. N. Cal.* 33.
NWC.

A. pilosa Labill., *Sert. Austro-Caledon.*: 12, t.17 (1824); *Fl. N. Cal.* 33.
NWC.

A. ramosa R. Br., *Prodr.*: 173 (1810); *Fl. Fiji* 319. *A. aspera* Swallen, *J. Wash. Acad. Sci.* 36: 177 (1936).
FIJ, NWC, TUB, VAN; Australia, New Zealand.

A. repens Trin., *Mém. Acad. Imp. Sci. St.-Pétersbourg, Sér. 6, Sci. Math.* 1: 87 (1831).
MRQ; Galapágos.

ERAGROSTIDEAE

45. Distichlis

D. spicata (L.) Greene, *Bull. Calif. Acad. Sci.* 2: 415 (1887). *Uniola spicata* L., *Sp. Pl.* 1: 71 (1753).
HAW; North & South America.

46. Eragrostis

1. Plants rhizomatous (rhizomes sometimes short and knotted); plants restricted to Hawaii:
  2. Glumes and sometimes lemmas long-ciliate .... **E. fosbergii**
  2. Glumes and lemmas not long-ciliate:
    3. Lower glume ≥ 3 mm long, apex exceeding lowermost lemma:
      4. Lemma apex obtuse and erose; anthers 0.7–1.2 mm long ...... **E. atropioides**
      4. Lemma apex acute; anthers 0.6–0.7 mm long .... **E. leptophylla**
    3. Lower glume ≤ 3 mm long, apex shorter than lowermost lemma:
      5. Spikelets mostly tightly clustered along branches; panicle branches ascending to erect, the lower branches mostly less than 6 cm long .... **E. variabilis**
      5. Spikelets more evenly spaced on branches; panicle branches mostly diverging, the lower more than 6.5 cm long .... **E. grandis**
1. Plants lacking rhizomes; plants mostly not just from Hawaii:
  6. Apex of lower glume extending beyond apex of first lemma:
    7. Plants annual .... **E. tef**
    7. Plants perennial:
      8. Spikelets c. 5 mm long; culms mostly 30 cm or less .... **E. monticola**
      8. Spikelets mostly more than 5 mm long; mature culms usually longer than 30 cm:
        9. Panicle axis moderately to densely scabrous-puberulous .... **E. leptophylla**
        9. Panicle axis glabrous:
          10. Pedicels of spikelets 5(– 8) mm or less; anthers 0.7–1 mm long .... **E. deflexa**

10. Pedicels of some spikelets 15 mm or longer; anthers 1–1.6 mm long .... **E. trichodes**

6. Apex of lower glume not extending to apex of first lemma:
    11. Spikelets < 3 mm long:
        12. Palea keel prominently ciliate, cilia at least 0.3–0.5 mm long:
            13. Pedicels mostly (or all) straight and shorter than the spikelets; anthers 2; plants 10–75 cm tall .... **E. ciliaris**
            13. Pedicels mostly curving and longer than the spikelets; anthers 3; plants mostly less than 40 cm tall .... **E. tenella**
        12. Palea keel merely scabrid:
            14. Ligules 0.4–0.6 mm long; plants eglandular; lemma 0.9–1.2 mm long ... . **E. japonica**
            14. Ligules 0.1–0.3 mm long; sheaths or culms occasionally with a few glandular depressions; lemma 1.2–1.8(– 2) mm long .... **E. pilosa**
    11. Spikelets > 3 mm long:
        15. Some mature spikelets with 35 or more florets:
            16. Culms of mature plants mostly 20 cm or shorter .... **E. paupera**
            16. Culms of mature plants longer than 20 cm (sometimes decumbent to prostrate):
                17. Spikelet terete or only slightly compressed laterally, its central axis frequently curving distally .... **E. dielsii**
                17. Spikelet laterally compressed, its central axis not curved:
                    18. Plants annual; spikelets ovate .... **E. unioloides**
                    18. Plants perennial; spikelets linear:
                        19. Leaf-sheaths typically villous at throat; lower glume 2–3 mm long; anthers 1–1.3 mm long .... **E. variabilis**
                        19. Leaf-sheaths glabrous; lower glume 0.9–1.5 mm long; anthers 0.3–0.7 mm long .... **E. spartinoides**
        15. Most mature spikelets with fewer than 30 florets:
            20. Spikelet falling entire; upper glume 3–4.5 mm long .... **E. superba**
            20. Spikelets breaking apart at maturity; upper glume usually ≤3 mm long (except sometimes in *E. elliottii*):
                21. Plants with glands on culms, leaves (margins or sheaths) or inflorescence (pedicels, rhachis, lemma keel):
                    22. Glands on culm lacking .... **E. parviflora**
                    22. Glands on culm present as depressions or annular rings:
                        23. Glands on culms occasional as depressions; glands on leaf margins absent .... **E. pilosa**
                        23. Glands on culms oblong or as annular rings; glands on leaf margins present or not:
                            24. Glands on pedicels present; anthers 0.3–0.8 mm long .... **E. leptostachya**
                            24. Glands on pedicels absent; anthers 0.2–0.4 mm long:
                                25. Sheaths eglandular; lemma keels glandular distally; anthers 0.3–0.4 mm long .... **E. cilianensis**

25. Sheaths sometimes glandular on midveins; lemma keels eglandular; anthers 0.2–0.3 mm long .... **E. minor**

21. Plants lacking glands:
    26. Anthers ≤ 0.5 mm long:
        27. Palea keels ciliate:
            28. Plants annual; lower glume ≤ 1.2 mm long .... **E. ciliaris**
            28. Plants perennial; lower glume ≥ 1.2 mm long .. **E. elongata**
        27. Palea keels glabrate to scabrous:
            29. Spikelets ovate; plants annual .... **E. unioloides**
            29. Spikelets more or less linear; plants annual or perennial:
                30. Upper glume 0.6–1 mm long; lower glume 0.6–1 mm long; lemma 0.9–1.2 mm long; anthers 0.1–0.2 mm long .... **E. japonica**
                30. Upper glume ≥ 1 mm long; lower glume 0.5–2.5 mm long; lemmas 1–3 mm long; anthers longer than 0.2 mm:
                    31. Plants perennial:
                        32. Panicle branches ≤ 4 cm long; spikelets strongly overlapping .... **E. spartinoides**
                        32. Panicle branches mostly > 4 cm; spikelets often distant or only one per branch:
                            33. Base of culms mostly decumbent; some pedicels 1–2 mm long ..... **E. brownii**
                            33. Base of culms erect; pedicels mostly 10–35 mm long .... **E. elliottii**
                    31. Plants annual:
                        34. Panicle branches usually whorled at lower 2 nodes; lemmatal veins inconspicuous .... **E. pilosa**
                        34. Panicle branches solitary or paired at the lower 2 nodes; lemmatal veins inconspicuous or moderately conspicuous:
                            35. Lemma 1.6–3 mm long; caryopsis 0.7–1.3 mm long, obovoid, whitish to light brown .... **E. tef**
                            35. Lemma 1–2.2 mm long; caryopsis 0.5–1.1 mm long, pyriform, brownish .... **E. pectinacea**
    26. Anthers > 0.5 mm long:
        36. Basal leaf-sheaths silky pubescent below (sometimes scantily) .... **E. curvula**
        36. Basal leaf-sheaths glabrous or sparsely hairy below:
            37. Palea deciduous from rhachilla (gently probe at oldest spikelets with teasing needle):

38. Anthers 2, 0.4–0.6 mm long; lemma apex acute to obtuse; pedicels 0.3–6 mm long; spikelets narrowly lanceolate . . . . **E. bahiensis**

38. Anthers 3, 0.7–0.9 mm long; lemma apex acute; pedicels 1–10 mm long; spikelets ovate-lanceolate . . . . **E. atrovirens**

37. Palea persistent on rhachilla:

39. Lower glume 0.5–1 mm long; upper glume 0.8–1.7 mm long; anthers 3 . . . . **E. tenuifolia**

39. Lower glume 1–3.4 mm long; upper glume 1.5–3.4 mm long; anthers 2 or 3:

40. Culms mostly decumbent (at least at base); inflorescence greatly exserted, lower half usually naked . . . . **E. brownii**

40. Culms mostly ascending to erect; panicle exsertion and branches variable:

41. Pedicels mostly or all shorter than spikelets; lemma midnerve prominently scabrous along length . . . . **E. scabriflora**

41. Pedicels mostly longer than spikelets; lemma midnerve obscurely scabrous near apex:

42. Leaf-blade scabrous on margin; plants 30–40 cm tall; panicle longer than wide; anthers 3, 0.6–1 mm long; caryopsis c. 1.2 mm long . . . . **E. mauiensis**

42. Leaf-blade smooth on margin; plants 40–80 cm tall; panicle length and width about the same; anthers 2, 0.3–0.8 mm long; caryopsis 0.6–0.8 mm long . . **E. elliottii**

E. atropioides Hillebr., *Fl. Hawaiian Isl.*: 531 (1888); *Man. Haw.* 1540.
HAW.

E. atrovirens (Desf.) Steud., *Nomencl. Bot.*, ed. 2, 1: 562 (1840); *List Micro.* 42. *Poa atrovirens* Desf., *Fl. Atlant.* 1: 73 (1798).
CRL; Old World tropics.

E. bahiensis Schrad. ex Schult., *Mant.* 2: 318 (1824); *List Micro.* 42
CRL; southern USA, South America. Voucher unknown, additional confirmation needed.

E. brownii (Kunth) Nees in Wight, *Cat. Indian Pl.*: 105 (1834); *List Micro.* 42; *Man. Haw.* 1540. *Poa brownii* Kunth, *Révis. Gramin.* 1: 112 (1829). *Eragrostis atrovirens* f. *brownii* (Kunth) Hack., *Allg. Bot. Z. Syst.* 21: 38 (1915); *Fl. N. Cal.* 35. *E. molokaiensis* H. St.John, *Phytologia* 64: 177 (1988).
CRL, EAS, HAW, MRN, MRS, NWC; Southeast Asia, Australia.

E. cilianensis (All.) Vignolo ex Janch., *Mitt. Naturwiss. Vereins Univ. Wien* 5(9): 110 (1907); *Man. Haw.* 1541. *Poa cilianensis* All., *Fl. Pedem.* 2: 246 (1785).
HAW, NWC; tropical and warm temperate regions.

E. ciliaris (L.) R. Br. in Tuckey, *Narr. Exped. Zaire*, app. 5: 478 (1818); *List Micro.* 42. *Poa ciliaris* L., *Syst. Nat.*, ed. 10, 2: 875 (1759).
CRL, HAW, MRN, MRS; tropics.

E. curvula (Schrad.) Nees, *Fl. Afr. Austral.* Ill.: 397 (1841). *Poa curvula* Schrad., *Gött. Gel. Anz.* 3: 2073 (1821).
HAW; South Africa and widely introduced as a cover plant.

E. deflexa Hitchc., *Mem. Bernice Pauahi Bishop Mus.* 8: 131 (1922); *Man. Haw.* 1541.
HAW.

E. dielsii Pilg. ex Diels & E. Pritz., *Bot. Jahrb. Syst.* 35: 76 (1904).
HAW; Australia.

E. elliottii S. Watson, *Proc. Amer. Acad. Arts* 25: 140 (1890).
CRL; Southeast USA and Central America.

E. elongata (Willd.) J.Jacq., *Ecl. Gram. Rar.*: 3, t. 3 (1813); *Fl. N. Cal.* 35; *Man. Haw.* 1538. *Poa elongata* Willd., *Enum. Pl.* 1: 108 (1809). *P. diandra* R. Br., *Prodr.*: 180 (1810). *Eragrostis diandra* (R. Br.) Steud., *Syn. Pl. Glumac.* 1: 279 (1854); *Fl. N. Cal.* 35.
HAW, NWC; Southeast Asia, Australia.

E. fosbergii Whitney, *Occas. Pap. Bernice Pauahi Bishop Mus.* 8: 75 (1937); *Man. Haw.* 1541.
HAW.

E. grandis Hillebr., *Fl. Hawaiian Isl.*: 528 (1888); *Man. Haw.* 1542. *E. grandis* var. *oligantha* Hillebr., l.c.: 528. *E. grandis* var. *polyantha* Hillebr., l.c.: 528. *E. polyantha* (Hillebr.) Jedwabn., *Bot. Arch.* 5: 200 (1924).
HAW.

E. japonica (Thunb.) Trin., *Gram. Gen.*: 405 (1830); *Fl. Sol.* 184. *Poa japonica* Thunb., *Fl. Jap.*: 51 (1784).
SOL; tropics.

E. leptophylla Hitchc., *Mem. Bernice Pauahi Bishop Mus.* 8: 133 (1922); *Man. Haw.* 1542.
HAW.

E. leptostachya (R. Br.) Steud., *Syn. Pl. Glumac.* 1: 279 (1854); *Easter* 77. *Poa leptostachya* R. Br., *Prodr.*: 180 (1810). *Eragrostis hosakai* O. Deg., *Fl. Hawaii., Fam.*: 47 (1940); *Man. Haw.* 1542.
EAS, HAW; Australia.

E. mauiensis Hitchc., *Mem. Bernice Pauahi Bishop Mus.* 8: 129 (1922); *Man. Haw.* 1543. *E. coerulea* Jedwabn, *Bot. Arch.* 5: 213 (1924).
HAW.

E. minor Host, *Icon. Descr. Gram. Austriac.* 4: 15 (1809); *Fl. N. Cal.* 35; *List Micro.* 42. *E. pooides* P. Beauv., *Ess. Agrostogr.*: 162 (1812); *Fl. Niue* 242.
MRN, NUE, NWC, WAK; warm temperate and subtropical Old World.

E. monticola (Gaudich.) Hillebr., *Fl. Hawaiian Isl.*: 531 (1888); *Man. Haw.* 1543. *Poa monticola* Gaudich., *Voy. Uranie*: 408 (1830).
HAW.

E. parviflora (R. Br.) Trin., *Mém. Acad. Imp. Sci. St.-Pétersbourg, Sér. 6, Sci. Math.* 1: 411 (1831); *Fl. Sol.* 184. *Poa parviflora* R. Br., *Prodr.*: 181 (1810). *Eragrostis depallens* Jedwabn., *Bot. Arch.* 5: 207 (1924). *E. novocaledonica* Jedwabn., l.c.: 207.
HAW, NWC, SOL; Indo-China, Australia.

E. paupera Jedwabn., *Bot. Arch.* 5: 214 (1924); *Man. Haw.* 1543. *E. whitneyi* Fosberg, *Occas. Pap. Bernice Pauahi Bishop Mus.* 15: 39 (1939); *List Micro.* 43. *E. whitneyi* var. *caumii* Fosberg, l.c.: 41.
GIL, HAW, LIN, PHX.

E. pectinacea (Michx.) Nees, *Fl. Afr. Austral. Ill.*: 406 (1841); *List Micro.* 42; *Man. Haw.* 1545. *Poa pectinacea* Michx., *Fl. Bor.-Amer.* 1: 69 (1803).
CRL, HAW; North and Central America.

E. pilosa (L.) P. Beauv., *Ess. Agrostogr.*: 71 (1812); *Fl. Poly.* 83; *Fl. N. Cal.* 35; *Fl. Fiji* 301; *List Micro.* 42; *Fl. Sol.* 184; *Man. Haw.* 1538; *Fl. Soc.* 336. *Poa pilosa* L., *Sp. Pl.* 1: 68 (1753). *Eragrostis petersii* Trin., *Mém. Acad. Imp. Sci. St.-Pétersbourg, Sér. 6, Sci. Math., Seconde Pt. Sci. Nat.* 1: 70 (1836).
CRL, FIJ, HAW, MRN, NWC, SCI, SOL; warm temperate and tropical Old World.

E. scabriflora Swallen, *J. Wash. Acad. Sci.* 26: 179 (1936); *Fl. Fiji* 303; *List. Micro.* 43. *E. tenax* Jedwabn., *Bot. Arch.* 5: 193 (1924), non Steud. (1840).
EAS, FIJ, LIN, MRS, NWC, VAN.

E. spartinoides Steud., *Syn. Pl. Glumac.* 1: 265 (1854); *Easter* 78. *E. carolinensis* Jedwabn., *Bot. Arch.* 5: 198 (1924).
CRL, EAS, LIN, NWC; Thailand, Australia.

E. superba Peyr., *Sitzungsber. Kaiserl. Akad. Wiss., Math.-Naturwiss. Cl.* 38: 584 (1860).
HAW; tropical Africa.

E. tef (Zucc.) Trotter, *Boll. Soc. Bot. Ital.* 1918: 62 (1918); *Fl. N. Cal.* 35. *Poa tef* Zucc., *Diss. Concern. Ist. Pianta Paniz. Abiss.* (1775).
NWC; a crop plant from Ethiopia.

E. tenella (L.) P. Beauv. ex Roem. & Schult., *Syst. Veg.* 2: 576 (1817); *Fl. Guam* 190; *Fl. Niue* 242; *Pl. Samoa* 30; *Fl. Fann.* 352; *Fl. Fiji* 303; *Fl. Sol.* 184; *Man. Haw.* 1545; *Fl. Soc.* 337. *Poa tenella* L., *Sp. Pl.* 1: 69 (1753). *Poa amabilis* auct. non L.; *P. plumosa* Retz., *Observ. Bot.* 4: 20 (1786). *Eragrostis amabilis* auct. non (L.) Wight & Arn.; *Fl. Poly.* 83; *Fl. Raro.* 19; *Pl. Tonga* 52; *List Micro.* 41. *E. elytroblephara* Steud., *Syn. Pl. Glumac.* 1: 280 (1854). *E. amabilis* var. *plumosa* (Retz.) A. Camus in Lecomte, *Fl. Indo-Chine* 7: 557 (1922); *Fl. N. Cal.* 35. *E. tenella* var. *insularis* C. E. Hubb., *Bull. Misc. Inform., Kew* 1939: 654 (1939).
COO, CRL, FIJ, GIL, HAW, LIN, MRN, MRQ, MRS, NRU, NUE, NWC, PHX, SAM, SCI, SOL, TON, TUA, TUB, WAK; tropics.

E. tenuifolia (A. Rich.) Hochst. ex Steud., *Syn. Pl. Glumac.* 1: 268 (1854). *Poa tenuifolia* A. Rich., *Tent. Fl. Abyss.* 2: 425 (1850).
COO, HAW, NWC; tropics.

E. trichodes (Nutt.) Wood, *Class-book Bot.*, 1861 ed.: 796 (1861). *Poa trichodes* Nutt., *Trans. Amer. Philos. Soc.* 5: 146 (1837).
HAW; southern USA.

E. unioloides (Retz.) Steud., *Syn. Pl. Glumac.* 1: 264 (1854); *Fl. Fiji* 301; *Man. Haw.* 1538. *Poa unioloides* Retz., *Observ. Bot.* 5: 19 (1789).
CRL, FIJ, HAW, NWC, SOL; tropical Asia.

E. variabilis (Gaudich.) Steud., *Nomencl. Bot.*, ed. 2, 1: 564 (1840); *Man. Haw.* 1545. *Poa variabilis* Gaudich., *Voy Uranie*: 408 (1829). *Eragrostis equitans* Trin., *Mém. Acad. Imp. Sci. St.-Pétersbourg, Sér. 6, Sci. Math.* 1: 413 (1831). *E. wahowensis* Trin., l.c.: 413. *E. hawaiiensis* Hillebr., *Fl. Hawaiian Isl.*: 530 (1888). *E. phleoides* Hillebr., l.c. 530. *E. thyrsoidea* Hillebr., l.c.: 529. *E. variabilis* var. *ciliata* Hillebr., l.c.: 529. *E. niihauensis* Whitney, *Occas. Pap. Bernice Pauahi Bishop Mus.* 13(5): 6 (1936). *E. hobdyi* H. St.John, *Phytologia* 64: 177 (1988).
HAW.

Note: Many species intergrade, at least slightly, and we recommend consulting local keys as well.

47. Ectrosiopsis

E. lasioclada (Merr.) Jansen, *Reinwardtia* 2: 268 (1953). *Eragrostis lasioclada* Merr., *Philipp. J. Sci. 1, Suppl.*: 382 (1906). *E. subaristata* Chase, *J. Arnold Arbor.* 20: 305 (1939). *Ectrosiopsis subaristata* (Chase) Jansen, *Reinwardtia* 2: 269 (1953); *List Micro.* 41. *Ectrosia lasioclada* (Merr.) S. T. Blake, *Proc. Roy. Soc. Queensland* 84: 65 (1973).
CRL; Australia.

48. Ectrosia

E. agrostoides Benth., *Fl. Austral.* 7: 634 (1878). *E. leporina sensu List Micro.* 41.
CRL; Moluccas, Australia.

49. Leptochloa

1. Plants annual:
   2. Ligules (0.5 –)1–2 mm long, apically erose; spikelets 1.9–2.5 mm long; lemma 0.8–1.2 mm long; lemma keel pubescent .... **L. panicea**
   2. Ligules 4–8 mm long, apically attenuate but becoming lacerate; spikelets 3–8 mm long; lemma 2–2.5 mm long; lemma keel glabrous .... **L. fusca**
1. Plants perennial:
   3. Leaf-blades filiform, stiff, convolute, 0.5–1 mm wide .... **L. xerophila**
   3. Leaf-blades herbaceous, 3–10 mm wide:
      4. Spikelets broadly linear, lightly compressed, loosely spaced in slender flexuous racemes; lemmas awnless .... **L. decipiens**
      4. Spikelets oblong, strongly compressed, overlapping; lemmas often awned:
         5. Racemes flexuous, the longer 6–12 cm .... **L. virgata**
         5. Racemes straight, stiff, the longer 3–4 cm .... **L. marquisensis**

L. decipiens (R. Br.) Stapf ex Maiden, *Agric. Gaz. New South Wales* 20: 307 (1909). *Poa decipiens* R. Br., *Prodr*.: 181 (1810). *Leptochloa capillacea sensu Fl. N. Cal*. 34.
NWC; Australia.
Note. Pacific specimens are all *L. decipiens* subsp. *decipiens*.

L. fusca (L.) Kunth, *Révis. Gramin*. 1: 91 (1829). *Poa malabarica* L., *Sp. Pl*. 1: 69 (1753), nom. rejic. *Festuca fusca* L., *Sp. Pl*., ed. 2, 1: 109 (1762). *Megastachya uninervia* J. Presl in C. Presl, *Reliq. Haenk*. 1: 279 (1830). *Centotheca malabarica* (L.) Merr., *Philipp. J. Sci*. 1, Suppl.: 385 (1906); *Fl. N. Cal*. 35. *Leptochloa uninervia* (J. Presl) Hitchc. & Chase, *Contr. US Natl. Herb*. 18: 383 (1917); *Man. Haw*. 1558.
HAW, NWC; tropics.
Note. Pacific specimens are all *Leptochloa fusca* subsp. *uninervia* (J. Presl) N. Snow, *Novon* 8: 79 (1998).

L. marquisensis (F. Br.) P. M. Peterson & Judz., *Taxon* 39: 659 (1990). *Eragrostis marquisensis* F. Br., *Bernice Pauahi Bishop Mus. Bull*. 84: 81 (1931).
MRQ.

L. panicea (Retz.) Ohwi, *Bot. Mag. (Tokyo)* 55: 311 (1941); *List Micro*. 47. *Poa panicea* Retz., *Observ. Bot*. 3: 11 (1783). *Eleusine filiformis* Pers., *Syn. Pl*. 1: 87 (1805). *Leptochloa filiformis* (Pers.) P. Beauv., *Ess. Agrostogr*.: 71 (1812); *List Micro*. 47.
MRN; tropics.
Note. Pacific specimens are all *Leptochloa panicea* subsp. *panicea*.

L. virgata (L.) P. Beauv., *Ess. Agrostogr*.: 71 (1812). *Cynosurus virgatus* L., *Syst. Nat*., ed. 10, 2: 876 (1759).
HAW; North & South America, introduced to Borneo & New Guinea.

L. xerophila P .M. Peterson & Judz., *Taxon* 39: 659 (1990). *Eragrostis xerophila* F. Br., *Bernice Pauahi Bishop Mus. Bull*. 84: 82 (1931, non Domin (1912).
MRQ.

50. Eleusine

1. Racemes 3–7 mm wide; spikelets elliptic, breaking up; grain elliptic to oblong, never exposed .... **E. indica**
1. Racemes 9–15 mm wide; spikelets ovate, persistent on inflorescence; grain almost globose, often exposed when mature .... **E. coracana**

E. coracana (L.) Gaertn., *Fruct. Sem. Pl.* 1: 8 (1788); *Fl. Fiji* 305; *List Micro.* 41. *Cynosurus coracanus* L., *Syst. Nat.*, ed. 10, 2: 875 (1759).
FIJ, MRN; Finger Millet, cultivated in the Old World tropics

E. indica (L.) Gaertn., *Fruct. Sem. Pl.* 1: 8 (1788); *Fl. Poly.* 85; *Fl. Raro.* 19; *Fl. N. Cal.* 34; *Pl. Tonga* 55; *Fl. Guam* 191; *Fl. Niue* 242; *Pl. Samoa* 62; *Fl. Fann.* 352; *Fl. Fiji* 304; *Fl. Pit.* 23; *List Micro.* 41; *Fl. Sol.* 184; *Man. Haw.* 1537; *Easter* 77; *Fl. Soc.* 335. *Cynosurus indicus* L., *Sp. Pl.* 1: 72 (1753).
COO, CRL, EAS, FIJ, GIL, HAW, LIN, MCS, MRN, MRQ, MRS, NRU, NUE, NWC, PHX, PIT, SAM, SCI, SOL, TON, TUA, TUB, VAN, WAK; throughout the tropics.

51. Dactyloctenium

D. aegyptium (L.) Willd., *Enum. Pl.*: 1029 (1809); *Fl. Poly.* 86; *Fl. Raro.* 18; *Pl. Tonga* 55; *Fl. Guam* 191; *Fl. Fiji* 306; *List Micro.* 35; *Fl. Sol.* 184; *Man. Haw.* 1521; *Fl. Soc.* 333. *Cynosurus aegyptius* L., *Sp. Pl.* 1: 72 (1753). *Dactyloctenium aegyptium* var. *radicans* Balansa, *Bull. Soc. Bot. France* 19: 318 (1872), nom. nud.; *Fl. N. Cal.* 34.
COO, CRL, FIJ, GIL, HAW, LIN, MCS, MRQ, MRN, MRS, NRU, NWC, SAM, SCI, SOL, TON, TUA, TUB, VAN, WAK; throughout the tropics.

52. Sporobolus

1. Panicle branches, at least the lower, in whorls; upper glume as long as spikelet:
   2. Panicle open and pyramidal at maturity; spikelets 1.6–1.8 mm long ... **S. pyramidatus**
   2. Panicle spiciform, linear; spikelets 1.8–2.5 mm long .... **S. piliferus**
1. Panicle branches not whorled:
   3. Plants annual; panicle oblong, diffuse, delicate; spikelets tiny, 0.8–1 mm long .... **S. tenuissimus**
   3. Plants perennial:
      4. Leaf-blades coriaceous, stiff; panicle spiciform, sometimes untidily so; spikelets 1.5–2.5 mm long:
         5. Plants densely tufted, without rhizomes; lower glume as long as spikelets; stamens 2 .... **S. farinosus**
         5. Plants rhizomatous, sward-forming; lower glume $^{2}/_{3}$–$^{4}/_{5}$ length of spikelets; leaves distichous; stamens 3 .... **S. virginicus**
      4. Leaf-blades herbaceous, flat when fresh:
         6. Stamens 2:
            7. Panicle open; spikelets 1.2–1.6 mm long .... **S. diandrus**

7. Panicle branches 0.5–2 cm long, distant, appressed to the axis, densely crowded with dark green spikelets 1.5–2 mm long .... **S. elongatus**

6. Stamens 3:

8. Panicle narrowly pyramidal; branches slender, flexuous, 5–15 cm long, spreading horizontally at anthesis; upper glume obtuse, $^1/_4$–$^1/_2$ as long as spikelet .... **S. pyramidalis**

8. Panicle linear, contracted to spiciform; branches stiff, mostly 1–2 cm long (longer in *S. fertilis*), ascending or appressed to the axis:

9. Lemma and palea scarcely longer than top of grain at maturity, not gaping; spikelets 1.5–2 mm long; upper glume acute .... **S. indicus**

9. Lemma and palea exceeding the grain by up to its own length, gaping open:

10. Spikelets 2.1–2.5 mm long; grain 1.1–1.2 mm long; upper glume acute .... **S. africanus**

10. Spikelets 1.7–2 mm long; grain 0.8–1 mm long; upper glume obtuse or acute .... **S. fertilis**

S. africanus (Poir.) A. Robyns & Tournay, *Bull. Jard. Bot. État* 25: 242 (1955); *Fl. Niue* 252; *List Micro.* 60; *Man. Haw.* 1596; *Easter* 82; *Fl. Soc.* 345. *Agrostis africana* Poir., *Encycl. Suppl.* 1: 254 (1810). *Sporobolus indicus* var. *africanus* (Poir.) Jovet & Guédès, *Taxon* 22: 163 (1973).
COO, EAS, HAW, MRN, NUE, SCI; tropical Africa, Australia, New Zealand.

S. diandrus (Retz.) P. Beauv., *Ess. Agrostogr.*: 26 (1812); *Fl. Guam* 193; *Fl. Fiji* 308; *List Micro.* 60; *Fl. Sol.* 186; *Man. Haw.* 1596; *Fl. Soc.* 346. *Agrostis diandra* Retz., *Observ. Bot.* 5: 19 (1789). *Sporobolus indicus* var. *diandrus* (Retz.) Jovet & Guédès, *Taxon* 22: 163 (1973).
CRL, FIJ, GIL, HAW, MRN, MRQ, MRS, NWC, SAM, SCI, SOL, TON, TUA, TUB, VAN; tropical Asia.

S. elongatus R. Br., *Prodr.*: 170 (1810); *Fl. Guam* 193; *Fl. Fiji* 308; *Fl. Pit.* 24; *Fl. Sol.* 186.
FIJ, HAW, MRN, NWC, PIT, SOL; Australia.

S. farinosus Hosok., *J. Soc. Trop. Agric.* 7: 321 (1935); *List Micro.* 61.
CRL, MRN.

S. fertilis (Steud.) Clayton, *Kew Bull.* 19: 291 (1965); *List Micro.* 61. *Agrostis fertilis* Steud., *Syn. Pl. Glumac.* 1: 170 (1854). *Sporobolus indicus* var. *fertilis* (Steud.) Jovet & Guédès, *Taxon* 22: 163 (1973). *S. indicus sensu Fl. Raro.* 21; *Pl. Tonga* 54; *Fl. Sol.* 186; *Man. Haw.* 1597; *Fl. Soc.* 346.
COO, CRL, FIJ, GIL, HAW, MRN, NUE, NWC, SCI, TON, TUB; Asia, Australia.

S. indicus (L.) R. Br., *Prodr.*: 170 (1810); *Fl. N. Cal.* 32; *Fl. Fiji* 307. *Agrostis indica* L., *Sp. Pl.* 1: 63 (1753).
FIJ, HAW, NWC; Central & South America.

S. piliferus (Trin.) Kunth, *Enum. Pl.* 1: 211 (1833). *Vilfa pilifera* Trin., *Gram. Unifl. Sesquifl.*: 157 (1824).
HAW; tropics.

S. pyramidalis P. Beauv., *Fl. Oware* 2: 36 (1816); *Fl. Sol.* 186. *S. jacquemontii* Kunth, *Révis. Gramin.* 2: 427 (1831); *Fl. Fiji* 307.
CRL, FIJ, MRN, MRQ, SOL; Africa, South America.

S. pyramidatus (Lam.) Hitchc., *Man. Grasses W. Ind.*: 84 (1936); *Man. Haw.* 1596. *Agrostis pyramidata* Lam., *Tabl. Encycl.* 1: 161 (1791).
HAW; southern USA to Argentina.

S. tenuissimus (Schrank) Kuntze, *Revis. Gen. Pl.* 3(2): 369 (1898). *Panicum tenuissimum* Schrank, *Denkschr. Königl.-Baier. Bot. Ges. Regensburg* 2: 26 (1822).
MRQ, SCI, TUB; tropics.

S. virginicus (L.) Kunth, *Révis. Gramin.* 1: 67 (1829); *Fl. N. Cal.* 32; *Fl. Guam* 192; *Fl. Fiji* 327; *List Micro.* 61; *Man. Haw.* 1597. *Agrostis virginica* L., *Sp. Pl.* 1: 63 (1753).
CRL, FIJ, HAW, MRN, MRS, NWC; throughout the tropics and subtropics.

Note. The taxonomy of the *Sporobolus indicus* complex (including *S. africanus* and *S. fertilis*) attempts to express observed differences between geographical populations, but the boundaries are very indistinct — a familiar feature of this genus. Various attempts to treat the taxa at infraspecific level have been made, but no concensus has yet emerged.

53. Muhlenbergia

M. microsperma (DC.) Trin., *Gram. Unifl. Sesquifl.*: 193 (1824); *Man. Haw.* 1562. *Trichochloa microsperma* DC., *Cat. Pl. Horti Monsp.*: 151 (1813).
HAW; California to Peru.

CYNODONTEAE

54. Chloris

1. Sterile florets inflated, obtriangular, as long as wide, two of them awned .... **C. barbata**
1. Sterile florets longer than wide, only one of them awned:
   2. Sterile lemma rudimentary, linear, 0.4–0.7 mm long .... **C. radiata**
   2. Sterile lemma well developed, oblong to cuneate:
      3. Sterile lemma bilobed for $^1/_3$–$^1/_2$ its length .... **C. divaricata**
      3. Sterile lemma not or inconspicuously bilobed:
         4. Fertile lemma with a dense plume of hairs 2–3 mm long at the apex **C. virgata**
         4. Fertile lemma with ciliate margins but no apical plume:
            5. Sterile florets 2–3 in number .... **C. gayana**
            5. Sterile floret 1, truncate .... **C. truncata**

C. barbata Sw., *Fl. Ind. Occid.* 1: 200 (1797); *Fl. Guam* 195; *Fl. Sol.* 184; *Man. Haw.* 1514; *Fl. Soc.* 331. *Andropogon barbatus* L., *Mant. Pl.* 2: 302 (1771), non L. (1759). *Chloris inflata* Link, *Enum. Hort. Berol. Alt.* 1: 105 (1821); *Fl. Fiji* 311; *List Micro.* 34.

CRL, FIJ, GIL, HAW, MRN, MRQ, MRS, NRU, NUE, NWC, PHX, SAM, SCI, SOL, TON, TUA, VAN, WAK; throughout the tropics.

C. divaricata R. Br., *Prodr.*: 186 (1810); *Man. Haw.* 1514. *C. cynodontoides* Balansa, *Bull. Soc. Bot. France* 19: 318 (1872); *Fl. N. Cal.* 34. *C. divaricata* var. *cynodontoides* (Balansa) Lazarides, *Austral. J. Bot., Suppl. Ser.* 5: 18 (1972); *Fl. Fiji* 312.
COO, FIJ, HAW, MRQ, NWC, TON; Australia.

C. gayana Kunth, *Révis. Gramin.* 1: 293 (1830); *Fl. N. Cal.* 34; *Fl. Guam* 195; *Fl. Niue* 237; *Fl. Fiji* 310; *List Micro.* 34; *Fl. Sol.* 184; *Man. Haw.* 1515; *Easter* 74.
EAS, FIJ, HAW, MRN, MRQ, MRS, NUE, NWC, TON; Africa and widely introduced as a fodder crop.

C. radiata (L.) Sw., *Prodr.*: 26 (1788); *Fl. Guam* 196; *List Micro.* 34; *Man. Haw.* 1515. *Agrostis radiata* L., *Syst. Nat.*, ed. 10, 2: 873 (1759).
CRL, HAW, MRN, MRQ, MRS, TON; Mexico to Brazil.

C. truncata R. Br., *Prodr.*: 186 (1810); *Fl. Fiji* 311; *Man. Haw.* 1513.
FIJ, HAW, NUE, TON; Australia, New Zealand.

C. virgata Sw., *Fl. Ind. Occid.* 1: 203 (1797); *List Micro.* 34; *Man. Haw.* 1515.
GIL, HAW, MRN, MRQ, NRU, NWC, SCI; tropics and subtropics.

55. Eustachys

E. petraea (Sw.) Desv., *Nouv. Bull. Sci. Soc. Philom. Paris* 2: 189 (1810); *List Micro.* 43; *Man. Haw.* 1513. *Chloris petraea* Sw., *Prodr.*: 25 (1788).
CRL, GIL, HAW, MRN, MRS; southern USA to Paraguay.

56. Enteropogon

1. Racemes 4–10, digitate, drooping; leaf-blades linear, 6–10 mm wide . . **E. dolichostachyus**
1. Racemes 1 (– 4), ascending; leaf-blades filiform, 0.5–1 mm wide . . . . **E. unispiceus**

E. dolichostachyus (Lag.) Keng ex Lazarides, *Austral. J. Bot., Suppl. Ser.* 5: 31 (1972). *Chloris dolichostachya* Lag., *Gen. Sp. Pl.*: 5 (1816). *C. incompleta* Roth ex Roem. & Schult., *Syst. Veg.* 2: 607 (1817); *List Micro.* 34.
CRL; tropical Asia, Australia.

E. unispiceus (F. Muell.) Clayton, *Kew Bull.* 21: 108 (1967). *Chloris unispicea* F. Muell., *Fragm.* 7: 118 (1870). *C. cheesemanii* Hack. ex Cheeseman, *Trans. Linn. Soc. London, Ser.* 2, *Bot.* 6: 305 (1903).
COO; Australia.

57. Cynodon

1. Plant with stolons and underground rhizomes . . . . **C. dactylon**
1. Plant with stolons but no rhizomes:
   2. Culms robust, woody; racemes in 2–5 whorls (rarely 1), stiff, red or purple . . . . **C. aethiopicus**
   2. Culms soft, not woody; racemes in 1 whorl (occasionally 2), slender, green or pigmented . . . . **C. nlemfuensis**

C. aethiopicus Clayton & J. R. Harlan, *Kew Bull.* 24: 187 (1970).
HAW; eastern Africa.

C. dactylon (L.) Pers., *Syn. Pl.* 1: 85 (1805); *Fl. Poly.* 87; *Fl. Raro.* 18; *Fl. N. Cal.* 34; *Pl. Tonga* 55; *Fl. Guam* 194; *Fl. Niue* 238; *Fl. Fann.* 352; *Fl. Fiji* 309; *List Micro.* 35; *Fl. Sol.* 184; *Man. Haw.* 1520; *Easter* 75; *Fl. Soc.* 332. *Panicum dactylon* L., *Sp. Pl.* 1: 58 (1753). *Cynodon maritimus* Kunth in Humb. & Bonpl., *Nov. Gen. Sp.* 1: 170 (1815). *C. dactylon* var. *maritimus* (Kunth) Hack., *Ark. Bot.* 8: 40 (1909); *List Micro.* 35. *C. parviglumis* Ohwi, *Bot. Mag. (Tokyo)* 55: 538 (1941). *C. dactylon* var. *parviglumis* (Ohwi) Fosberg & Sachet, *Micronesica* 18: 45 (1984); *List Micro.* 35.
COO, CRL, EAS, FIJ, GIL, HAW, HBI, LIN, MRN, MRQ, MRS, NRU, NUE, NWC, PHX, PIT, SAM, SCI, SOL, TON, TUA, TUB, WAK; throughout tropical and warm temperate regions, where commonly used as a lawn grass.

C. nlemfuensis Vanderyst, *Bull. Agric Congo Belge* 13: 342 (1922); *Man. Haw.* 1520.
HAW, NUE; tropical Africa.

58. Tragus

1. Spikelets subequal, both fertile, scarcely separated by an internode much shorter than the basal stipe . . . . **T. australianus**
1. Spikelets unequal, the upper narrower and sterile, separated by a distinct internode almost as long as the basal stipe of the pair . . . . **T. berteronianus**

T. australianus S. T. Blake, *Pap. Dept. Biol. Univ. Queensland* 1(18): 12 (1941). *T. racemosus sensu Fl. N. Cal.* 31.
NWC; Australia.

T. berteronianus *Schult.*, *Mant.* 2: 205 (1824); *Man. Haw.* 1601.
HAW; tropical Africa, southern USA to Argentina.

59. Zoysia

1. Spikelets ovate, 2–2.5 times as long as wide; pedicels flexuous, usually longer than spikelets . . . . **Z. japonica**
1. Spikelets lanceolate to oblong, 3–4 times as long as wide; pedicels straight, usually shorter than spikelets . . . . **Z. matrella**

Z. japonica Steud., *Syn. Pl. Glumac.* 1: 414 (1854); *Fl. Guam* 197; *Fl. Fiji* 319; *List Micro.* 63. *Z. matrella* var. *japonica* (Steud.) Sasaki, *List Pl. Formosa*: 80 (1928). *Z. matrella* subsp. *japonica* (Steud.) Masam. & Yanagita, *Trans. Nat. Hist. Soc. Taiwan* 31: 327 (1941).
FIJ, MRN; northern China, Taiwan, Japan.

Z. matrella (L.) Merr., *Philipp. J. Sci., Bot.* 7: 230 (1912); *Fl. Guam.* 197; *List Micro.* 63; *Fl. Sol.* 186; *Fl. Soc.* 348. *Agrostis matrella* L., *Mant. Pl.* 2: 185 (1771). *Zoysia pungens* Willd., *Ges. Naturf. Freunde Berlin Neue Schriften* 3: 441 (1801); *Fl. Poly.* 51. *Z. tenuifolia* Thiele, *Linnaea* 9: 309 (1835); *Fl. Guam* 197; *Fl. Sol.* 186. *Z. matrella* var. *pacifica* Goudsw., *Blumea* 26: 172 (1980); *List Micro.* 63. *Z. pacifica* (Goudsw.) M. Hotta & Kuroki, *Acta Phytotax. Geobot.* 45: 71 (1994).
CRL, FIJ, HAW, MRN, MRQ, MRS, SAM, SCI, SOL, TUB; tropical Asia and widely introduced as a lawn grass. A narrow-leaved form is sometimes treated as *Zoysia matrella* var. *pacifica*.

60. Lepturopetium

1. Spikelets 4.5–6 mm long with 2 fertile florets; glumes smooth; apical lemma lanceolate with an awn 1–3 mm long . . . . **L. kuniense**
1. Spikelets 3–3.5 mm long with 1 fertile floret; glumes scabrid; apical lemma linear, awnless . . . . **L. marshallense**

L. kuniense Morat, *Adansonia* 20: 378 (1981).
NWC.

L. marshallense Fosberg & Sachet, *Micronesica* 18: 72 (1984); *List Micro.* 47.
COO, MRN, MRS.

61. Lepturus

L. repens (G. Forst.) R. Br., *Prodr.*: 207 (1810); *Fl. Poly.* 88; *Fl. N. Cal.* 32; *Pl. Tonga* 54; *Fl. Guam* 196; *Pl. Samoa* 87; *Fl. Fann.* 352; *Fl. Fiji* 313; *Fl. Pit.* 23; *List Micro.* 48; *Fl. Sol.* 185; *Man. Haw.* 1560; *Easter* 78; *Fl. Soc.* 338. *Rottboellia repens* G. Forst., *Fl. Ins. Austr.*: 9 (1786). *Lepturus acutiglumis* Steud., *Syn. Pl. Glumac.* 1: 357 (1854); *Fl. Fiji* 313; *Fl. Soc.* 338. *L. repens* var. *maldenensis* F. Br., *Occas Pap. Bernice Pauahi Bishop Mus.* 9: 6 (1930); *Fl. Poly.* 89. *L. repens* var. *palmyrensis* F. Br., l.c.: 6; *Fl. Poly.* 89. *L. cinereus* Burcham, *Contr. U.S. Natl. Herb.* 30: 424 (1948); *Fl. Niue* 244. *L. mildbraedianus* I. Hansen & Potztal, *Bot. Jahrb. Syst.* 76: 269 (1954). *L. pilgerianus* I. Hansen & Potztal, l.c.: 268; *List Micro.* 48. *L. repens* var. *cinereus* (Burcham) Fosberg, *Occas. Pap. Bernice Pauahi Bishop Mus.* 21: 292 (1954); *List Micro.* 48. *L. repens* var. *occidentalis* Fosberg, l.c.: 291; *List Micro.* 48. *L. repens* var. *septentrionalis* Fosberg, l.c.: 291; *List Micro.* 48. *L. repens* var. *subulatus* Fosberg, l.c.: 290; *List Micro.* 48. *L. gasparricensis* Fosberg, l.c.: 293; *List Micro.* 47.
COO, CRL, EAS, FIJ, GIL, HAW, HBI, LIN, MCS, MRN, MRS, NRU, NUE, NWC, PHX, PIT, SAM, SCI, SOL, TON, TUA, TUB, TUV, VAN, WAK; sea shores from eastern Africa to Australia.

PANICEAE

62. Oplismenus

1. Awns scaberulous .... **O. burmannii**
1. Awns smooth and viscid:
   2. Spikelets distant, adjacent pairs on lower racemes 4–12 mm apart; lowest raceme 2.5–10 cm long .... **O. compositus**
   2. Spikelets contiguous, adjacent pairs on lower racemes 0.5–4 mm apart; lowest raceme 0.5–3 cm long .... **O. hirtellus**

O. burmannii (Retz.) P. Beauv., *Ess. Agrostogr.*: 54, 169 (1812); *Pl. Tonga* 5. *Panicum burmannii* Retz., *Observ. Bot.* 3: 10 (1783).
HAW, TON; tropics.

O. compositus (L.) P. Beauv., *Ess. Agrostogr.*: 54, 169 (1812); *Fl. Poly.* 67; *Fl. Raro.* 19; *Fl. N. Cal.* 29; *Pl. Tonga* 61; *Fl. Guam* 214; *Fl. Niue* 247; *Fl. Pit.* 23; *List Micro.* 50; *Fl. Sol.* 185; *Man. Haw.* 1563; *Fl. Soc.* 340. *Panicum compositum* L., *Sp. Pl.* 1: 57 (1753). *P. setarium* Lam., *Tab. Encycl.* 1: 170 (1791). *Oplismenus setarius* (Lam.) Roem. & Schult., *Syst. Veg.* 2: 481 (1817); *Fl. Poly.* 68; *Fl. N. Cal.* 29. *O. compositus* var. *setarius* (Lam.) F. M. Bailey, *Queensland Grass.*: 19 (1888). *O. patens* Honda, *Repert. Spec. Nov. Regni Veg.* 20: 360 (1924). *O. compositus* f. *glabratus* F. Br., *Bernice P. Bishop Mus. Bull.* 84: 68 (1931). *O. compositus* f. *pubescens* F. Br., l.c.: 68. *O. setarius* f. *sterilis* F. Br., l.c.: 68. *O. compositus* var. *patens* (Honda) Ohwi, *Acta Phytotax. Geobot.* 11: 35 (1942); *List Micro.* 50.
COO, CRL, FIJ, HAW, MRN, MRQ, MRS, NUE, NWC, PIT, SAM, SCI, SOL, TON, TUA, TUB, VAN; tropics but principally Asia.

O. hirtellus (L.) P. Beauv., *Ess.Agrostogr.*: 54, 170 (1812); *Fl. N. Cal.* 29; *Fl. Niue* 247; *Fl. Pit.* 23; *Man. Haw.* 1563. *Panicum hirtellum* L., *Syst. Nat.*, ed. 10, 2: 870 (1759). *Orthopogon imbecillis* R. Br., *Prodr.*: 194 (1810). *Oplismenus imbecillis* (R. Br.) Roem. & Schult., *Syst. Veg.* 2: 487 (1817); *Pl. Tonga* 61; *Fl. Fiji* 342; *Fl. Soc.* 340. *Panicum oahuaense* Steud., *Nomencl. Bot.*, ed. 2, 2: 260 (1841), nom. nud. *Oplismenus oahuaensis* Steud., l.c.: 220, in syn. *O. undulatifolius* var. *imbecillis* (R. Br.) Hack., *Publ. Bur. Sci. Gov. Lab.* 25: 82 (1905); *Fl. Sol.* 185. *O. microphyllus* Honda, *J. Fac. Sci. Univ. Tokyo, Sect. 3, Bot.* 3: 274 (1930). *O. hirtellus* subsp. *imbecillis* (R. Br.) U. Scholz, *Phanerog. Monogr.* 13: 127 (1981). *O. hirtellus* var. *imbecillis* (R. Br.) Fosberg & Sachet, *Micronesica* 18: 78 (1984); *List Micro.* 50. *O. hirtellus* var. *microphyllus* (Honda) Fosberg & Sachet, l.c.: 79; *List Micro.* 50. *O. undulatifolius* sensu *Fl. Raro.* 19; *Fl. Guam* 214.
COO, CRL, FIJ, HAW, MRN, MRQ, NUE, NWC, PIT, SAM, SCI, SOL, TON, TUB, VAN; tropics.

63. Panicum

1. Upper lemma rugose .... **P. maximum**
1. Upper lemma smooth or faintly striate:
   2. Spikelets hairy:

3. Spikelets densely villous:
 4. Leaf-blades velvety pubescent; annual .... **P. torridum**
 4. Leaf-blades glabrous, or rarely pilose above:
  5. Plants annual .... **P. ramosius**
  5. Plants perennial .... **P. beecheyi**
3. Spikelets loosely hairy:
 6. Spikelets gibbous, 1–1.5 mm long, puberulous; leaf-blades lanceolate to ovate; lower glume $^1/_4$–$^1/_2$ length of spikelet; panicle effuse .... **P. trichoides**
 6. Spikelets symmetrical, 1.5–4.5 mm long; leaf-blades linear; lower glume $^3/_4$ to as long as spikelet:
  7. Spikelets pubescent, usually with a tuft of apical hairs .... **P. fauriei**
  7. Spikelets tuberculate-hirsute:
   8. Spikelets acute, with 0.4–1.5 mm long hairs; lower glume shorter than spikelet .... **P. pellitum**
   8. Spikelets acuminate, with 1–3 mm long hairs; lower glume as long as spikelet .... **P. konaense**
2. Spikelets glabrous:
 9. Lower glume $^4/_5$ to as long as spikelet:
  10. Plants annual:
   11. Leaf-sheaths puberulous; panicle linear with pubescent appressed branches; leaf-blades loosely involute .... **P. fauriei**
   11. Leaf-sheaths tuberculate-pilose; panicle lanceolate to ovate with glabrous or sparsely pilose ascending branches; leaf-blades flat ..... **P. xerophilum**
  10. Plants perennial:
   12. Leaf-blades 1–7 mm wide, more or less involute .... **P. tenuifolium**
   12. Leaf-blades 8–25 mm wide, flat:
    13. Panicle branches spreading, the spikelets loose ..... **P. nephelophilum**
    13. Panicle branches appressed, the spikelets densely clustered .... **P. niihauense**
 9. Lower glume $^1/_{10}$–$^3/_4$ length of spikelet:
  14. Leaves mostly basal, forming a low mat or cushion; panicle of few broadly elliptic spikelets:
   15. Leaf-blade margins tuberculate-ciliate throughout; blades lanceolate to ovate, 3.5–8 mm wide .... **P. isachnoides**
   15. Leaf-blade margins eciliate (or with few basal cilia on those immediately below panicle — *P. hillebrandianum*):
    16. Leaf-blades 4–8 mm wide, lanceolate .... **P. hillebrandianum**
    16. Leaf-blades 1–3 mm wide:
     17. Top of peduncle and adjacent panicle axis pubescent to pilose; leaf-blades narrowly lanceolate, 1–3 mm wide .... **P. cynodon**
     17. Top of peduncle and panicle axis glabrous or almost so; leaf-blades filiform to linear, 0.7–1.8 mm wide .... **P. koolauense**
  14. Leaves basal and cauline:
   18. Plants annual:
    19. Spikelets 4.5–5 mm long, persistent on panicle; lower glume $^1/_2$–$^3/_4$ length of spikelet .... **P. miliaceum**

19. Spikelets 1.5–3 mm long, shed separately or with the whole panicle:
    20. Spikelets acuminate, falling with the intact panicle; lower palea absent or up to $^1/_2$ the length of its lemma . . . . **P. capillare**
    20. Spikelets acute, deciduous; lower palea as long as its lemma . . . . **P. luzonense**

18. Plants perennial:
    21. Lower glume oblate, clasping, $^1/_4$–$^1/_3$ length of spikelet, truncate to obtuse, rarely acute:
        22. Culm soft, spongy, prostrate and rooting at the nodes below; leaf-blades 8–14 mm wide, floating . . . . **P. paludosum**
        22. Culm firm:
            23. Plant rhizomatous; leaf-blades distichous, stiff, often involute . . . . **P. repens**
            23. Plant tufted, without rhizomes:
                24. Panicle branches stiff, straight, the spikelets clustered towards the tip; lower floret barren . . . . **P. decompositum**
                24. Panicle branches flexuous, the spikelets evenly distributed:
                    25. Leaf-blades linear, herbaceous; lower floret male with palea; glumes separated by a brief internode 0.1–0.2 mm long . . . . **P. coloratum**
                    25. Leaf-blades lanceolate, coriaceous, stiff; lower floret barren without palea; glumes arising at almost the same level . . . . **P. longivaginatum**
    21. Lower glume ovate, $^1/_2$–$^3/_4$ length of spikelet, acute to acuminate:
        26. Spikelets 2mm long; panicle effuse . . . . **P. palauense**
        26. Spikelets 2.4–5 mm long:
            27. Spikelets acute, not gaping, 2.4–3.2 mm long; margins of upper glume hyaline; lower floret barren, its lemma 7–9-nerved; spikelets more or less clustered about primary branches . . . . **P. antidotale**
            27. Spikelets acuminate, gaping, 3–5 mm long; margins of upper glume firm . . . . **P. lineale**

P. antidotale Retz., *Observ. Bot*. 4: 17 (1786); *Fl. Fiji* 346; *List Micro*. 51; *Man. Haw*. 1567.
FIJ, HAW, MRN; Middle East, tropical Asia, Australia.

P. beecheyi Hook. & Arn., *Bot. Beechey Voy*.: 100 (1841); *Man. Haw*. 1561. *P. kauaiense* Hitchc., *Mem. Bernice Pauahi Bishop Mus*. 8: 187 (1922). *P. lihauense* H. St.John, *Phytologia* 63: 370 (1987). *P. mokuleiaense* H. St.John, l.c.: 371.
HAW.

P. capillare L., *Sp. Pl*. 1: 58 (1753); *Fl. N. Cal*. 30.
NWC; North & South America and widely introduced.

P. coloratum L., *Mant. Pl*. 1: 30 (1767); *Fl. Fiji* 346; *Fl. Sol*. 185; *Man. Haw*. 1567.
FIJ, HAW, SOL; Africa.

P. cynodon Reichardt, *Sitzungber. Kaiserl. Akad. Wiss., Math.-Naturwiss. Cl., Abt.* I 76: 724 (1878). *P. imbricatum* Hillebr., *Fl. Hawaiian Isl.*: 501 (1888). *P. forbesii* Hitchc., *Mem. Bernice Pauahi Bishop Mus.* 8: 199 (1922). *P. imbricatum* var. *oreoboloides* Whitney, *Occas. Pap. Bernice Pauahi Bishop Mus.* 13: 173 (1937). *P. alakaiense* Skottsb., *Acta Horti Gothob.* 25: 294 (1944). *P. imbricatum* f. *minus* Skottsb., l.c.: 250. *P. imbricatum* f. *molokaiense* Skottsb., l.c.: 290. *P. oreoboloides* (Whitney) Skottsb., l.c.: 292. *P. oreoboloides* var. *subimbricatum* Skottsb., l.c.: 292. *P. lamiatile* H. St.John, *Pacific Sci.* 25: 39 (1971). *P. lustriale* H. St.John, l.c.: 41. *Dichanthelium cynodon* (Reichardt) C. A. Clark & Gould, *Brittonia* 30: 58 (1978); *Man. Haw.* 1525. *D. forbesii* (Hitchc.) C. A. Clark & Gould, l.c.: 58. *Panicum baltodes* H. St.John, *Phytologia* 63: 368 (1987). *P. hobdyi* H. St.John, l.c.: 369. *P. pepeopaeense* H. St.John, l.c.: 371.
HAW.

P. decompositum R. Br., *Prodr.*: 191 (1810); *Fl. N. Cal.* 30; *Pl. Tonga* 60. *P. amabilis* Balansa, *Bull. Soc. Bot. France* 19: 324 (1872).
FIJ, NWC, TON; Australia.

P. fauriei Hitchc., *Mem. Bernice Pauahi Bishop Mus.* 8: 182 (1922); *Man. Haw.* 1568. *P. carteri* Hosaka, *Occas. Pap. Bernice Pauahi Bishop Mus.* 17: 6 (1942). *P. degeneri* Potztal, *Mitt. Bot. Gart. Berlin-Dahlem* 1: 128 (1953). *P. moomomiense* H. St.John, *Rhodora* 78(815): 542 (1976). *P. nubigenum* var. *latius* H. St.John, *Phytologia* 47: 376 (1981). *P. annuale* H. St.John, *Phytologia* 63: 368 (1987). *P. kukaiwaaense* H. St.John, l.c.: 370. *P. malikoense* H. St.John, l.c.: 371. *P. ninoleense* H. St.John, l.c.: 371. *P. sylvanum* H. St.John, l.c.: 372. *P. fauriei* var. *carteri* (Hosaka) Davidse, *Ann. Missouri Bot. Gard.* 77: 589 (1990). *P. fauriei* var. *latius* (H. St.John) Davidse, l.c.: 589.
HAW.

P. hillebrandianum Hitchc., *Mem. Bernice Pauahi Bishop Mus.* 8: 197 (1922). *P. monticola* Hillebr., *Fl. Hawaiian Isl.*: 501 (1888), non Hook. f. (1864). *P. conjugens* Skottsb., *Acta Horti Gothob.* 15: 298 (1944). *P. hillebrandianum* var. *gracilius* Skottsb., l.c.: 296. *P. gracilius* (Skottsb.) H. St.John, *Phytologia* 36: 465 (1977). *Dichanthelium conjugens* (Skottsb.) C. A. Clark & Gould, *Brittonia* 30: 57 (1978). *D. hillebrandianum* (Hitchc.) C. A. Clark & Gould, l.c.:57; *Man. Haw.* 1526. *Panicum infraventale* H. St.John, *Phytologia* 63: 369 (1987). *P. kahiliense* H. St.John, l.c.: 369. *P. knudsenii* H. St.John, l.c.: 370. *P. kokeense* H. St.John, l.c. 370. *P. ooense* H. St.John, l.c.: 371. *P. wilburii* H. St.John, l.c.: 372.
HAW.

P. isachnoides Hillebr., *Fl. Hawaiian Isl.*: 501 (1888). *P. isachnoides* var. *kilohanae* Skottsb., *Acta Horti Gothob.* 15: 288 (1944). *Dichanthelium isachnoides* (Hillebr.) C. A. Clark & Gould, *Brittonia* 30: 57 (1978); *Man. Haw.* 1526.
HAW.

P. konaense Whitney & Hosaka, *Occas. Pap. Bernice Pauahi Bishop Mus.* 12(5): 3 (1936); *Man. Haw.* 1568. *P. cookei* H. St.John, *Phytologia* 63: 368 (1987). *P. kanaioense* H. St.John, l.c.: 370. *P. waikoloaense* H. St.John, l.c.: 372.
HAW.

P. koolauense H. St.John & Hosaka, *Occas. Pap. Bernice Pauahi Bishop Mus.* 11(13): 3 (1935). *Dichanthelium koolauense* (H. St.John & Hosaka) C. A. Clark & Gould, *Brittonia* 30: 58 (1978); *Man. Haw.* 1526. *Panicum ekeanum* H. St.John, *Phytologia* 63: 369 (1987). *P. waimeaense* H. St.John, l.c.: 372.
HAW.

P. lineale H. St.John, *Phytologia* 63: 370 (1987); *Man. Haw.* 1569.
HAW.

P. longivaginatum H. St.John, *Phytologia* 63: 370 (1987).
HAW.

P. luzonense J. Presl in C. Presl, *Reliq. Haenk.* 1: 308 (1830); *Fl. Guam* 215; *List Micro.* 52.
CRL, MRN, SOL; Southeast Asia.

P. maximum Jacq., *Icon. Pl. Rar.* 1: 2, t.13 (1781); *Fl. Poly.* 70; *Fl. Raro.* 19; *Fl. N. Cal.* 30; *Pl. Tonga* 60; *Fl. Guam* 213; *Fl. Niue* 247; *Fl. Fiji* 345; *List Micro.* 52; *Fl. Sol.* 185; *Man. Haw.* 1569; *Easter* 79; *Fl. Soc.* 341. *P. maximum* var. *trichoglume* Robyns, *Mém. Inst. Roy. Colon. Belge, Sect. Sci. Nat.* 1(6): 31 (1932); *Fl. Fiji* 340. *Urochloa maxima* (Jacq.) R. D. Webster, *Austral. Paniceae*: 241 (1987). *Megathyrsus maximus* (Jacq.) B. K. Simon & S. W. L. Jacobs, *Austrobaileya* 6: 572 (2003).
COO, CRL, EAS, FIJ, HAW, MRN, MRO, MRS, NUE, NWC, SAM, SCI, SOL, TON, TUA, TUB, VAN; Africa and introduced throughout the tropics.

P. miliaceum L., *Sp. Pl.* 1: 58 (1753); *Man. Haw.* 1570.
CRL, HAW, MRN; Proso millet, grown as a grain crop, particularly in Asia.

P. nephelophilum Gaudich., *Voy. Uranie*: 411 (1829); *Man. Haw.* 1570. *P. havaiense* Reichardt, *Sitzungber. Kaiserl. Akad. Wiss., Math.-Naturwiss. Cl., Abt.* 1, 76(1): 723 (1878). *P. kaalaense* Hitchc., *Mem. Puauhi Bishop Mus.* 8: 191 (1922). *P. honokowaiense* H. St.John, *Phytologia* 63: 369 (1987). *P. nephelophilum* var. *levius* H. St.John, l.c.: 371.
HAW.

P. niihauense H. St.John, *Occas. Pap. Bernice Puauhi Bishop Mus.* 9(14): 5 (1931); *Man. Haw.* 1570.
HAW.

P. palauense Ohwi, *Bot. Mag. (Tokyo)* 55: 544 (1941).
CRL. No voucher seen.

P. paludosum Roxb., *Fl. Ind. ed.* 1820, 1: 310 (1820). *P. telmatodes* Balansa, *Bull. Soc. Bot. France* 19: 324 (1872); *Fl. N. Cal.* 30.
NWC, VAN; tropical Asia, Australia.

P. pellitum Trin., *Gram. Panic.*: 198 (1826); *Man. Haw.* 1571. *P. pseudagrostis* Trin., l.c.: 197. *P. colliei* Endl., *Ann. Wiener Mus. Naturgesch.* 1: 157, No. 571 (1836), nom. nud. *P. affine*

Hook. & Arn., *Bot. Beechey Voy.*: 100 (1832), non Poir. (1816). *P. gossypinum* Hook. & Arn., l.c.: 100. *P. lanaiense* Hitchc., *Mem. Bernice Puauhi Bishop Mus.* 8: 129 (1922). *P. pellitoides* F. Br. & H. St.John, *Occas. Pap. Bernice Puauhi Bishop Mus.* 10(12): 3 (1934).
HAW.

P. ramosius Hitchc., *J. Bot.* 71: 6 (1933); *Man. Haw.* 1571.
HAW.

P. repens L., *Sp. Pl.* ed. 2, 1: 87 (1762); *List Micro.* 52; *Man. Haw.* 1571.
CRL, HAW, MRN; Old World tropics.

P. tenuifolium Hook. & Arn., *Bot. Beechey Voy.*: 100 (1832); *Man. Haw.* 1573. *P. nephelophilum* var. *rhyacophilum* Hillebr., *Fl. Hawaiian Isl.*: 498 (1988). *P. nephelophilum* var. *tenuifolium* (Hook. & Arn.) Hillebr., l.c.: 497. *P. molokaiense* O. Deg. & Whitney in O. Deg., *Fl. Hawaii., Fam.* 47 (1936).
HAW.

P. torridum Gaudich., *Voy. Uranie*: 411 (1829); *Man. Haw.* 1573. *Neurachne montanum* Gaudich., l.c.: t. 26 (1827); *Panicum montanum* Gaudich., l.c.: 411 (1829), non Roxb. (1820). *Neurachne torrida* (Gaudich.) Kunth, *Enum. Pl.* 1: 114 (1833). *Panicum nubigenum* Kunth, l.c.: 98. *P. cinereum* Hillebr., *Fl. Hawaiian Isl.*: 500 (1888).
HAW.

P. trichoides Sw., *Prodr.*: 24 (1788); *Fl. Sol.* 185.
SOL; Central & South America.

P. xerophilum (Hillebr.) Hitchc., *Mem. Bernice Puauhi Bishop Mus.* 8: 193 (1922); *Man. Haw.* 1574. *P. nephelophilum* var. *xerophilum* Hillebr., *Fl. Hawaiian Isl.*: 498 (1888). *P. heupueo* H. St.John, *Pacific Sci.* 13: 156 (1959). *P. comae* H. St.John, *Phytologia* 47: 374 (1981). *P. assurgens* H. St.John, *Phytologia* 63: 368 (1987), non Renvoize (1982). *P. bifurcatum* H. St.John, l.c.: 368. *P. furtivum* H. St.John, l.c.: 369, non Swallen (1950). *P. kahoolawense* H. St.John, l.c.: 369. *P. kaonohuaense* H. St.John, l.c.: 370. *P. simplex* H. St.John, l.c.: 371, non Trin. (1836). *P. subglabrum* H. St.John, l.c.: 372.
HAW.

64. Ancistrachne

1. Spikelets 3.1–3.3 mm long .... **A. numaeensis**
1. Spikelets 4.3–5 mm long .... **A. uncinulata**

A. numaeensis (Balansa) S. T. Blake, *Proc. Roy. Soc. Queensland* 81: 1 (1969). *Panicum numaeense* Balansa, *Bull. Soc. Bot. France* 19: 325 (1872); *Fl. N. Cal.* 30.
NWC.

A. uncinulata (R. Br.) S. T. Blake, *Pap. Dept. Biol. Univ. Queensland* 1(19): 5 (1941); *Fl. Fiji* 361. *Panicum uncinulatum* R. Br., *Prodr.*: 181 (1810).
FIJ; Australia.

65. Sacciolepis

S. indica (L.) Chase, *Proc. Biol. Soc. Wash.* 1: 8 (1908); *Fl. N. Cal.* 30; *Fl. Guam* 224; *Fl. Fiji* 360; *List Micro.* 56; *Man. Haw.* 1589. *Aira indica* L., *Sp. Pl.* 1: 63 & Errata (1753).
CRL, FIJ, HAW, MRN, NWC, TUB; Old World tropics.

66. Cyrtococcum

1. Pedicels longer than spikelets; panicle loose, the branches ascending .... **C. patens**
1. Pedicels mostly shorter than spikelets; panicle contracted, the branches more or less appressed to the axis:
   2. Leaf-blades mostly 6–20 cm long; panicle branches usually pilose; spikelets usually glabrous .... **C. oxyphyllum**
   2. Leaf-blades mostly 1–4.5 cm long; panicle branches glabrous; spikelets usually hispidulous .... **C. trigonum**

C. oxyphyllum (Steud.) Stapf, *Hooker's Icon. Pl.* t. 3096 (1922); *Pl. Tonga* 60; *Fl. Niue* 238; *Fl. Fiji* 351; *List Micro.* 35; *Fl. Sol.* 184; *Fl. Soc.* 333. *Panicum oxyphyllum* Steud., *Syn. Pl. Glumac.* 1: 65 (1853).
COO, CRL, FIJ, NUE, NWC, SAM, SCI, SOL, TON, VAN, WAL; tropical Asia.

C. patens (L.) A. Camus, *Bull. Mus. Hist. Nat. (Paris)* 27: 118 (1921); *List Micro.* 35; *Fl. Sol.* 184. *Panicum patens* L., *Sp. Pl.* 1: 58 (1753); *Fl. N. Cal.* 30. *P. accrescens* Trin., *Sp. Gram.* 1: t. 8 (1828). *Cyrtococcum accrescens* (Trin.) Stapf, *Hooker's Icon Pl.*: t. 3096 (1922); *Fl. Sol.* 184.
CRL, MRN, NWC, SAM, SOL; tropical Asia.

C. trigonum (Retz.) A. Camus, *Bull. Mus. Hist. Nat. (Paris)* 27: 118 (1921); *Pl. Sam.* 110; *Fl. Fiji* 351. *Panicum trigonum* Retz., *Observ. Bot.* 3: 9 (1783); *Pl. Tonga* 60.
COO, FIJ, SAM, TON, VAN; tropical Asia.

67. Echinochloa

1. Ligule represented by a fringe of hairs .... **E. picta**
1. Ligule absent:
   2. Spikelets acuminate to awned, in 2–several irregular rows; longest raceme 2–10 cm long:
      3. Spikelets mostly 3–4 mm long; upper floret 2–3 mm long; lowest raceme often with secondary branchlets .... **E. crusgalli**
      3. Spikelets 3.8–6.5 mm long; upper floret 3.5–5 mm long; racemes all simple .... **E. oryzoides**
   2. Spikelets acute to cuspidate; racemes seldom over 3 cm long, simple:
      4. Racemes neatly 4-rowed, openly spaced, commonly about $^{1}/_{2}$ their length apart, appressed to the axis; spikelets 1.5–3 mm long; lower floret male or barren .... **E. colona**
      4. Racemes crowded with plump spikelets, congested into a dense lanceolate head; spikelets 2.5–4 mm long, purple tinged; lower floret barren ...... **E. esculenta**

E. colona (L.) Link, *Hort. Berol.* 2: 209 (1833); *Fl. Poly.* 69; *Fl. Raro.* 19; *Fl. N. Cal.* 29; *Pl. Tonga* 61; *Fl. Guam* 212; *Pl. Samoa* 110; *Fl. Fiji* 341; *List Micro.* 40; *Fl. Sol.* 84; *Man. Haw.* 1535; *Fl. Soc.* 335. *Panicum colonum* L., *Syst. Nat.*, ed. 10, 2: 870 (1759).
COO, CRL, FIJ, HAW, MRN, MRQ, NUE, NWC, SAM, SCI, SOL, TON, TUB, VAN, WAL; tropics and subtropics.

E. crusgalli (L.) P. Beauv., *Ess. Agrostogr.*: 53 (1812); *Fl. N. Cal.* 29; *Fl. Niue* 241; *Fl. Sol.* 184; *Man. Haw.* 1535; *Fl. Soc.* 335. *Panicum crusgalli* L., *Sp. Pl.* 1: 56 (1753). *Echinochloa crusgalli* var. *austrojaponensis* Ohwi, *Acta Phytotax. Geobot.* 11: 38 (1942); *List Micro.* 40. *E. glabrescens* Kossenko, *Bot. Mater. Gerb. Bot. Inst. Komarova Acad. Nauk SSSR* 11: 40 (1949); *List Micro.* 41. *E. cruspavonis* sensu *Fl. Sol.* 184.
CRL, FIJ, GIL, HAW, MRQ, MRS, NUE, NWC, SCI, SOL, TUB; warm temperate & subtropical regions.

E. esculenta (A. Braun) H. Scholz, *Taxon* 41: 523 (1992). *Panicum esculentum* A. Braun, *Ind. Sem. Berlin App.*: 3 (1861). *Echinochloa utilis* Ohwi & Yabuno, *Acta Phytotax. Geobot.* 20: 50 (1962). *E. frumentacea* subsp. *utilis* (Ohwi & Yabuno) Tzvelev, *Novosti Syst. Vyssh. Rast.* 1968: 17 (1968). *E. frumentacea sensu Fl. Fiji* 340.
FIJ, HAW; eastern Asia, Australia, cultivated as a grain crop.

E. oryzoides (Ard.) Fritsch, *Verh. Zool.-Bot. Ges. Wien* 41: 742 (1891). *Panicum oryzoides* Ard., *Animadv. Bot. Spec. Alt.* 2: 16, t.5 (1764). *P. hispidulum* Retz., *Observ. Bot.* 5: 18 (1789). *Echinochloa crusgalli* var. *hispidula* (Retz.) Honda, *Bot. Mag. (Tokyo)* 37: 122 (1923); *Fl. Fiji* 341; *List Micro.* 41.
CRL. FIJ, HAW; Mediterranean & Middle East, becoming a widespread weed of rice.

E. picta (J. Koenig) P. W. Michael, *Philipp. Weed J. Sci.* 5: 18 (1978); *List Micro.* 41. *Panicum pictum* J. Koenig, *Naturforscher (Halle)* 23: 204 (1788). *Echinochloa stagnina sensu Fl. Fiji* 340.
FIJ, HAW, MRN, SAM; tropical Asia.

68. Alloteropsis

A. semialata (R. Br.) Hitchc., *Contr. U.S. Natl. Herb.* 12: 210 (1909); *Fl. N. Cal.* 28. *Panicum semialatum* R. Br., *Prodr.*: 192 (1810).
NWC, SOL; Old World tropics.

69. Urochloa

1. Upper lemma with a distinct mucro 0.5–1.2 mm long; upper glume and lower lemma exceeding the upper floret and enclosing its mucro:
    2. Lower glume 3-veined, narrower than spikelet; spikelets cuspidate; upper lemma rugulose .... **U. mosambicensis**
    2. Lower glume 5-veined, as wide as the spikelet; spikelets acuminate; upper lemma striate .... **U. glumaris**
1. Upper lemma without a mucro (at most obscurely mucronulate in *U. mutica*); upper glume and lower lemma not or scarcely exceeding the upper floret:

3. Lower glume $^{3}/_{4}$ to as long as spikelet, 11-veined; spikelets 4–6 mm long .... **U. humidicola**
3. Lower glume up to $^{1}/_{2}$ length of spikelet:
   4. Margins of raceme rhachis tuberculate-ciliate; spikelets 4–6 mm long; glumes not separated by an internode:
      5. Raceme rhachis crescentic, solid, 1 mm wide; plant caespitose ... **U. brizantha**
      5. Raceme rhachis herbaceous, flat, winged, 1–1.7 mm wide; plant stoloniferous .. **U. decumbens**
   4. Margins of raceme rhachis eciliate, sometimes the surface pilose:
      6. Raceme rhachis triquetrous, without wings; spikelets 1.5–2.7 mm long:
         7. Upper lemma smooth, shiny, obtuse, readily shed from spikelet; lower glume up to $^{1}/_{5}$ length of spikelet .... **U. eruciformis**
         7. Upper lemma striate to rugose, acute or mucronulate, persistent:
            8. Lower glume up to $^{1}/_{4}$ length of spikelet, veinless; upper lemma rugose .. **U. reptans**
            8. Lower glume $^{1}/_{3}$–$^{1}/_{2}$ length of spikelet, 3–5-veined; upper lemma striate or obscurely rugulose .... **U. mollis**
      6. Raceme rhachis flat, ribbon-like, winged, 0.5–1.5 mm wide:
         9. Spikelets paired in several untidy rows, 2.5–3.5 mm long; glumes not separated by an internode; perennial .... **U. mutica**
         9. Spikelets borne singly in 1–2 rows; glumes separated by a short internode; annual:
            10. Length of spikelets 4–5.5 mm .... **U. plantaginea**
            10. Length of spikelets 2.4–3.7 mm:
               11. Spikelets 2.4–3 mm long; inflorescence axis usually 0.5–2 cm long .. **U. distachya**
               11. Spikelets 3.3–3.7 mm long; inflorescence axis usually 3–10 cm long .. **U. subquadripara**

U. brizantha (Hochst. ex A. Rich.) R. D. Webster, *Austral. Paniceae*: 233 (1987). *Panicum brizanthum* Hochst. ex A. Rich, *Tent. Fl. Abyss.* 2: 363 (1850). *Brachiaria brizantha* (Hochst. ex A. Rich.) Stapf in Prain, *Fl. Trop. Afr.* 9: 531 (1919); *Fl. Fiji* 332.
FIJ, HAW, SCI; Africa and introduced as a pasture grass.

U. decumbens (Stapf) R. D. Webster, *Austral. Paniceae*: 234 (1987). *Brachiaria decumbens* Stapf in Prain, *Fl. Trop. Afr.* 9: 528 (1919); *Fl. Sol.* 183.
HAW, SOL; Central Africa and introduced as a pasture grass.

U. distachya (L.) T. Q. Nguyen, *Novosti Sist. Vyssh. Rast.* 1966: 13 (1966). *Panicum distachyon* L., *Mant.* 1: 138 (1771); *Fl. Raro.* 19; *List Micro.* 51. *Brachiaria distachya* (L.) Stapf in Prain, *Fl. Trop. Afr.* 9: 565 (1919); *Fl. Guam* 204; *Fl. Sol.* 183.
COO, CRL, FIJ, GIL, HAW, MRN, MRS, SOL; tropical Asia.

U. eruciformis (Sm.) C. Nelson & Fern. Casas, *Fontqueria* 51: 4 (1998). *Panicum eruciforme* Sm., *Fl. Graec.* 1(2): 44, t. 59 (1808); *List Micro.* 51. *Echinochloa eruciformis* (Sm.) Rchb., *Fl. Germ. Excurs.* 3: 45 (1833); *Fl. N. Cal.* 29. *Brachiaria eruciformis* (Sm.) Griseb. in

Ledeb., *Fl. Ross.* 4: 469 (1853); *Fl. Guam* 204; *Fl. Fiji* 331. *Moorochloa eruciformis* (Sm.) Veldkamp, *Reinwardtia* 12: 139 (2004).
FIJ, MRN, MRS, NWC; eatern Africa to China.

U. glumaris (Trin.) Veldkamp, *Blumea* 41: 420 (1996). *Panicum glumare* Trin., *Gram. Panic.*: 143 (1826). *Urochloa paspaloides* J. Presl in C.Presl, *Reliq. Haenk.* 1: 318 (1830); *Fl. Soc.* 347. *Panicum ambiguum* Trin., *Mém. Acad. Imp. Sci. St.-Pétersbourg, Sér. 6, Sci. Math, Seconde Pt. Sci. Nat.* 3(2): 243 (1835), non. Turq. (1815); *Fl. Poly.* 71; *Fl. Raro.* 19; *Pl. Tonga* 59; *List Micro.* 51. *P. infidum* Steud., *Syn. Pl. Glumac.* 1: 63 (1853); *Fl. N. Cal.* 30. *Brachiaria ambigua* A. Camus in Lecomte, *Fl. Indo-Chine* 7: 433 (1922); *Fl. N. Cal.* 30. *B. paspaloides* (J. Presl) C. E. Hubb., *Hooker's Icon. Pl.* 34: t.3363 (1938); *Fl. Guam* 205; *Fl. Niue* 235; *Fl. Fiji* 330; *Fl. Sol.* 183. *Urochloa ambigua* (A. Camus) Pilg. in Engl. & Prantl, *Nat. Pflanzenfam., Aufl.* 2, 14e: 35 (1940).
COO, CRL, FIJ, MRN, MRS, NUE, NWC, SAM, SCI, SOL, TON, TUB, VAN; tropical Asia.

U. humidicola (Rendle) Morrone & Zuloaga, *Darwiniana* 31: 80 (1992). *Panicum humidicola* Rendle in Hiern, *Cat. Afr. Pl.* 2: 169 (1899). *Brachiaria humidicola* (Rendle) Schweick., *Bull. Misc. Inform., Kew* 1936: 297 (1936); *Fl. Fiji* 330; *Fl. Sol.* 183.
FIJ, SOL; tropical Africa.

U. mollis (Sw.) Morrone & Zuloaga, *Darwiniana* 31: 85 (1992). *Panicum molle* Sw., *Prodr.*: 22 (1788); *Fl. N. Cal.* 30. *Brachiaria mollis* (Sw.) Parodi, *Darwiniana* 15: 100 (1969).
HAW, NWC; Central & South America.

U. mosambicensis (Hack.) Dandy, *J. Bot.* 69: 54 (1931); *Fl. Sol.* 186. *Panicum mosambicense* Hack., *Bol. Soc. Brot.* 6: 140 (1888).
SOL; eastern & southern Africa.

U. mutica (Forssk.) T. Q. Nguyen, *Novosti Sist. Vyssh. Rast.* 1966: 13 (1966). *Panicum muticum* Forssk., *Fl. Aegypt.-Arab.*: 20 (1775); *List Micro.* 52. *Brachiaria mutica* (Forssk.) Stapf in Prain, *Fl. Trop. Afr.* 9: 526 (1919); *Fl. N. Cal.* 29; *Fl. Guam* 203; *Fl. Niue* 234; *Fl. Fiji* 329; *Fl. Sol.* 183; *Man. Haw.* 1504; *Fl. Soc.* 330.
CRL, FIJ, HAW, MRN, MRQ, MRS, NUE, NWC, SAM, SCI, SOL, TON; throughout the tropics.

U. plantaginea (Link) R. D. Webster, *Syst. Bot.* 13: 607 (1988). *Panicum plantagineum* Link, *Hort. Berol.* 1: 206 (1827). *Brachiaria plantaginea* (Link) Hitchc. in *Contr. U.S. Natl. Herb.* 12: 212 (1909); *Fl. Fann.* 352.
HAW, LIN; Central & South America.

U. reptans (L.) Stapf in Prain, *Fl. Trop. Afr.* 9: 601 (1920); *Fl. N. Cal.* 29. *Panicum reptans* L., *Syst. Nat.*, ed. 10, 2: 870 (1759); *List Micro.* 52. *P. prostratum* Lam., *Tab. Encycl.* 1: 171 (1791); *Fl. Soc.* 341. *P. taitense* Steud., *Syn. Pl. Glumac.* 1: 418 (1854). *P. patulum* Mez, *Notizbl. Bot. Gart. Berlin-Dahlem* 7: 64 (1917), non Hitchc. (1908). *P. prostratum* var. *marquisense* F. Br., *Bernice P. Bishop Mus. Bull.* 84: 71 (1931). *Brachiaria reptans* (L.) C. A. Gardner & C. E. Hubb., *Hooker's Icon. Pl.* 34: t.3363 (1938); *Fl. Guam* 203; *Fl. Fiji* 331; *Fl. Sol.* 183; *Fl. Soc.* 330.

CRL, FIJ, HAW, MRN, MRQ, MRS, NWC, SAM, SCI, SOL, TON; tropical Asia and widely introduced.

U. subquadripara (Trin.) R. D. Webster, *Austral. Paniceae*: 252 (1987). *Panicum subquadriparum* Trin., *Gram. Panic.*: 145 (1826); *List Micro.* 53. *P. miliiforme* J. Presl in C. Presl, *Reliq. Haenk.* 1: 300 (1830); *List Micro.* 52. *Brachiaria miliiformis* (J. Presl) Chase, *Contr. U.S. Natl. Herb.* 32: 35 (1920); *Fl. Guam* 204; *Fl. Sol.* 183. *B. subquadripara* (Trin.) Hitchc., *Lingnan Sci.* J. 7: 214 (1931); *Pl. Tonga* 56; *Fl. Guam* 205; *Fl. Niue* 235; *Fl. Fiji* 332; *Fl. Sol.* 183; *Man. Haw.* 1503.
COO, CRL, FIJ, GIL, HAW, MRN, MRS, NUE, NWC, PHX, SAM, SOL, TON, VAN; tropical Asia, Australia.

Note. The genus is accepted here in a wide sense, but its generic boundaries are still somewhat fluid.

70. Eriochloa

1. Upper lemma with a mucro 0.3–0.5 mm long; annual .... **E. procera**
1. Upper lemma with an awn 1–1.5 mm long; perennial .... **E. punctata**

E. procera (Retz.) C. E. Hubb., *Bull. Misc. Inform., Kew* 1930: 256 (1930); *Fl. Fiji* 328; *List Micro.* 43. *Agrostis procera* Retz., *Observ. Bot.* 4: 19 (1786).
CRL, FIJ, HAW, MRN, SAM, TON; tropical Asia, Australia.

E. punctata (L.) Ham., *Prodr. Pl. Ind. Occid.*: 5 (1825); *Man. Haw.* 1546. *Milium punctatum* L., *Syst. Nat.*, ed. 10, 2: 872 (1759).
HAW; Central & South America.

71. Entolasia

E. marginata (R. Br.) Hughes, *Bull. Misc. Inform., Kew* 1923: 331 (1923). *Panicum marginatum* R. Br., *Prodr.*: 190 (1810).
HAW; New Guinea, Australia, New Zealand.

72. Thuarea

T. involuta (G. Forst.) Roem. & Schult., *Syst. Veg.* 2: 808 (1917); *Fl. Raro.* 22; *Fl. N. Cal.* 31; *Pl. Tonga* 63; *Fl. Guam* 226; *Pl. Samoa* 110; *Fl. Fiji* 354; *List Micro.* 62; *Fl. Sol.* 186; *Fl. Soc.* 347. *Ischaemum involutum* G. Forst., *Fl. Ins. Austr.*: 73 (1786). *Thuarea sarmentosa* Pers., *Syn. Pl.* 1: 110 (1805); *Fl. Poly.* 78.
COO, CRL, FIJ, GIL, HAW, MRN, MRS, NWC, PIT, SAM, SCI, SOL, TON, TUA, TUB, VAN; Southeast Asia.

73. Paspalum

1. Glumes both absent .... **P. malacophyllum**

1. Glumes, or at least the upper, present:
   2. Upper glume fringed with a ragged papery wing .... **P. fimbriatum**
   2. Upper glume wingless:
      3. Spikelets with a ciliate fringe from margin of upper glume, at least towards apex:
         4. Upper floret dark brown; spikelets elliptic to obovate, obtuse .... **P. virgatum**
         4. Upper floret yellow; spikelets elliptic, acute:
            5. Racemes paired; plant stoloniferous; spikelets yellow ...... **P. conjugatum**
            5. Racemes 3–20; plant tufted:
               6. Racemes mostly 3–7; spikelets 2.8–4 mm long .... **P. dilatatum**
               6. Racemes 10–20; spikelets 2–2.8 mm long .... **P. urvillei**
      3. Spikelets glabrous to pubescent but without a ciliate fringe:
         7. Upper floret brown:
            8. Spikelets borne singly in 2 rows .... **P. scrobiculatum**
            8. Spikelets paired in 4 rows .... **P. longifolium**
         7. Upper floret pallid or yellow:
            9. Spikelets borne singly:
               10. Leaf blades pilose on both sides; racemes 2–5; plant tufted .. **P. thunbergii**
               10. Leaf blades glabrous; racemes paired (rarely 3–5):
                  11. Plant rhizomatous; upper glume and lower lemma cartilaginous, glabrous; spikelets broadly elliptic, plumply plano-convex .... **P. notatum**
                  11. Plant stoloniferous:
                     12. Upper glume thinly coriaceous, obscurely puberulous; spikelets ovate, plump .... **P. distichum**
                     12. Upper glume papery, glabrous; spikelets ovate-elliptic, flattened .. .. **P. vaginatum**
            9. Spikelets paired:
               13. Upper glume puberulous:
                  14. Spikelets suborbicular, 1.3–1.4 mm long; racemes numerous, (10–) 15–60 .... **P. paniculatum**
                  14. Spikelets elliptic, 2–2.7 mm long; racemes few, 5–10(– 15) .... **P. macrophyllum**
               13. Upper glume and lower lemma glabrous:
                  15. Spikelets acute, elliptic, 2–3 mm long; racemes 4–6... **P. forsterianum**
                  15. Spikelets obtuse, orbicular or almost so, 1.4–2.6 mm long; racemes 1–4 .... **P. setaceum**

P. conjugatum P. J. Bergius, *Acta Helv. Phys.-Math.* 7: 129 (1772); *Fl. Poly.* 77; *Fl. Raro.* 20; *Fl. N. Cal.* 31; *Pl. Tonga* 58; *Fl. Guam* 216; *Fl. Niue* 248; *Pl. Samoa* 141; *Fl. Fiji* 336; *Fl. Pit.* 23; *List Micro.* 53; *Fl. Sol.* 185; *Man. Haw.* 1575; *Easter* 79; *Fl. Soc.* 341.
COO, CRL, EAS, FIJ, HAW, MRN, MRQ, MRS, NUE, NWC, PIT, SAM, SCI, SOL, TON, TUB, VAN; throughout the tropics.

P. dilatatum Poir. in Lam., *Encycl.* 5: 35 (1804); *Fl. Raro.* 20; *Fl. N. Cal.* 31; *Pl. Tonga* 58; *Fl. Guam* 216; *Fl. Niue* 248; *Fl. Fiji* 337; *List Micro.* 54; *Fl. Sol.* 185; *Man. Haw.* 1576; *Easter* 79.

COO, EAS, FIJ, HAW, MRN, NUE, NWC, SAM, SOL, TON; South America and widely introduced.

P. distichum L., *Syst. Nat.*, ed. 10, 2: 855 (1759); *Fl. Fiji* 336.
FIJ, HAW, NWC; warm temperate & subtropical regions.

P. fimbriatum Kunth in Humb. & Bonpl., *Nov. Gen. Sp.* 1: 93 (1815); *Fl. Guam* 216; *List Micro.* 54; *Fl. Sol.* 185; *Man. Haw.* 1576.
HAW, MRN, MRS, SOL, TON, VAN; Mexico to Brazil.

P. forsterianum Flüggé, *Gram. Monogr., Paspalum*: 172 (1810); *Easter* 79. *P. undulatum* Spreng. in Biehler, *Pl. Nov. Herb. Spreng.* 5: 6 (1807), non Poir. (1804). *P. venustum* P. Beauv., *Ess. Agrostogr.* 11, 172 (1812), nom. nud.
EAS, NWC.

P. longifolium Roxb., *Fl. Ind. ed.* 1820, 1: 283 (1820); *List Micro.* 54; *Fl. Sol.* 185. *P. orbiculare* var. *otobedii* Fosberg & Sachet, *Micronesica* 18: 83 (1984).
CRL, HAW, MRN, NRU, NWC, SOL; Southeast Asia, northern Australia.

P. macrophyllum Kunth in Humb. & Bonpl., *Nov. Gen. Sp.* 1: 92 (1815).
HAW; northern South America.

P. malacophyllum Trin., *Sp. Gram.* 3: t. 271 (1830).
HAW; South America.

P. notatum Flüggé, *Gram. Monogr., Paspalum*: 106 (1810); *Fl. Fiji* 337; *List Micro.* 54; *Fl. Sol.* 185; *Easter* 80.
EAS, FIJ, HAW, MRN, SOL; Central & South America.

P. paniculatum L., *Syst. Nat.*, ed. 10, 2: 855 (1759); *Fl. Poly.* 76; *Fl. N. Cal.* 31; *Fl. Fiji* 335; *List Micro.* 54; *Fl. Sol.* 185; *Man. Haw.* 1575; *Easter* 80; *Fl. Soc.* 342.
COO, CRL, EAS, FIJ, HAW, MRN, MRQ, NWC, SAM, SCI, SOL, TUB. VAN; South America and widely introduced.

P. scrobiculatum L., *Mant. Pl.* 1: 29 (1767); *Fl. N. Cal.* 31; *Fl. Guam* 217; *Man. Haw.* 1576; *Easter* 81; *Fl. Soc.* 342. *P. orbiculare* G. Forst., *Fl. Ins. Austr.*: 7 (1786); *Fl. Poly.* 77; *Fl. Raro.* 21; *Fl. N. Cal.* 31; *Pl. Tonga* 59; *Fl. Guam* 218; *Fl. Niue* 249; *Fl. Fiji* 338; *Fl. Pit.* 23; *List Micro.* 54; *Fl. Sol.* 185. *P. commersonii* Lam., *Tab. Encycl.* 1: 175, t.43/1 (1791); *Fl. Guam* 218; *List Micro.* 53. *P. cartilagineum* J. Presl in C. Presl, *Reliq. Haenk.* 1: 216 (1830); *Pl. Tonga* 58; *Fl. Guam* 217; *List Micro.* 53; *Fl. Sol.* 185. *P. cartilagineum* var. *biglumaceum* Fosberg & Sachet, *Micronesica* 18: 81 (1984); *List Micro.* 53. *P. moratii* Toutain, *Austrobaileya* 3: 724 (1992).
COO, CRL, EAS, FIJ, HAW, MRN, NRU, NWC, PIT, SAM, SCI, SOL, TON, TUA, TUB, VAN, WAK; Old World tropics.

P. setaceum Michx., *Fl. Bor.-Amer.* 1: 43 (1803). *P. ciliatifolium* Michx., l.c.: 44. *P. setaceum* var. *ciliatifolium* (Michx.) Vasey, *Contr. U.S. Natl. Herb.* 3: 17 (1892); *List Micro.* 55.
HAW, MRN, MRS, SAM; North America.

P. thunbergii Steud., *Syn. Pl. Glumac.* 1: 28 (1853); *Fl. Niue* 249.
NUE; Southeast Asia.

P. urvillei Steud., *Syn. Pl. Glumac.* 1: 24 (1853); *Fl. Guam* 217; *Fl. Fiji* 339; *List Micro.* 55; *Man. Haw.* 1576.
COO, FIJ, HAW, MRN, NWC, SAM; South America and widely introduced.

P. vaginatum Sw., *Prodr.*: 21 (1788); *Fl. N. Cal.* 31; *Pl. Tonga* 59; *Fl. Niue* 249; *Man. Haw.* 1577. *P. distichum sensu Fl. Poly.* 76; *Fl. Raro.* 20; *Fl. Guam* 217; *List Micro.* 54; *Fl. Soc.* 341.
COO, CRL, FIJ, HAW, MRQ, NUE, NWC, SAM, SCI, TOK, TON, TUA, TUB; throughout the tropics.

P. virgatum L., *Syst. Nat.*, ed. 10, 2: 855 (1759).
HAW; Central & South America.

74. Axonopus

1. Spikelets densely hairy .... **A. paschalis**
1. Spikelets glabrous or with sparse hairs not obscuring the veins:
   2. Upper lemma almost as long as spikelet; culm nodes glabrous .... **A. fissifolius**
   2. Upper lemma $^{4}/_{5}$ length of spikelet; culm nodes pubescent .... **A. compressus**

A. compressus (Sw.) P. Beauv., *Ess. Agrostogr.*: 12 (1812); *Fl. N. Cal.* 31; *Pl. Tonga* 58; *Fl. Guam* 202; *Fl. Niue* 232; *Pl. Samoa* 142; *Fl. Fiji* 333; *List Micro.* 31; *Fl. Sol.* 183; *Fl. Soc.* 329. *Milium compressum* Sw., *Prodr.*: 24 (1788).
CRL, FIJ, HAW, MRN, MRQ, NUE, NWC, SAM, SCI, SOL, TON; tropical America and widely introduced as a lawn grass.

A. fissifolius (Raddi) Kuhlm., *Comiss. Linhas. Telegr. Estrateg. Mato Grosso Amazonas* 11: 87 (1922); *Man. Haw.* 1500. *Paspalum fissifolium* Raddi, *Agrostogr. Bras.*: 26 (1823). *Axonopus affinis* Chase, *J. Wash. Acad. Sci.* 28: 180 (1938); *Fl. Fiji* 334; *Fl. Sol.* 183.
FIJ, HAW, MRQ, NWC, SAM, SCI, SOL, TUA; subtropical & tropical America.

A. paschalis (Stapf) Pilger in Skottsb., *Nat. Hist. Juan Fernandez* 2: 63 (1922); *Easter* 73. *Paspalum paschale* Stapf, *Bull. Misc. Inform., Kew* 1913: 117 (1913). *P. scoparium* var. *oligostachyum* Hack. ex Fuentes, *Inst. Cent. Meteor. Chile Publ.* 4: 145 (1913).
EAS.

75. Setaria

1. Panicle open or contracted, clearly branched:
   2. Leaf-blades pleated fanwise:
      3. Upper lemma strongly rugose; annual; leaf-blades 5–30 mm wide ..... **S. barbata**
      3. Upper lemma obscurely rugulose; perennial; leaf-blades 15–80 mm wide .... **S. palmifolia**

2. Leaf-blades flat; upper lemma smooth; lower glume acuminate and $^1/_2$–$^2/_3$ length of spikelet (if obtuse and $^1/_4$–$^1/_3$ length of spikelet see *Dissochondrus*, which occasionally has a herbaceous lower lemma):
   4. Callus of upper floret glabrous; culms ascending, branched; leaf-blades 9–15 cm long .... **S. austrocaledonica**
   4. Callus of upper floret pubescent; culms erect, unbranched; leaf-blades 25–30 cm long .... **S. jaffrei**
1. Panicle spiciform, sometimes lobed; leaf-blades flat:
   5. Spikelets persistent, eventually shedding their upper lemma, this almost smooth .... **S. italica**
   5. Spikelets deciduous as a whole; upper lemma rugose:
      6. Bristles retrorsely barbed, adhering tenaciously to clothing .... **S. verticillata**
      6. Bristles antrorsely barbed (species difficult to distinguish without basal parts):
         7. Plants annual .... **S. pumila**
         7. Plants perennial:
            8. Culms loosely ascending from a short knotty rhizome .... **S. parviflora**
            8. Culms densely caespitose, erect .... **S. sphacelata**

S. austrocaledonica (Balansa) A. Camus, *Bull. Mus. Natl. Hist. Nat.* 34: 181 (1928); *Fl. N. Cal.* 28. *Panicum austrocaledonicum* Balansa, *Bull. Soc. Bot. France* 19: 326 (1872).
NWC.

S. barbata (Lam.) Kunth, *Révis. Gramin.* 1: 47 (1829); *Fl. N. Cal.* 28; *Fl. Fiji* 348; *Fl. Sol.* 185. *Panicum barbatum* Lam., *Tab. Encycl.* 1: 171 (1791).
FIJ, NWC, SOL; Africa.

S. italica (L.) P .Beauv., *Ess. Agrostogr.*: 51 (1812). *Panicum italicum* L., *Sp. Pl.* 1: 56 (1753).
HAW; Foxtail Millet, widely cultivated in the tropics and subtropics.

S. jaffrei Morat, *Adansonia*, n.s. 18: 258 (1978).
NWC.

S. palmifolia (J. Koenig) Stapf, *J. Linn. Soc., Bot.* 42: 186 (1914); *Pl. Samoa* 143; *Fl. Fiji* 348; *Fl. Sol.* 186; *Man. Haw.* 1592. *Panicum palmifolium* J. Koenig, *Naturforscher* 23: 208 (1788).
FIJ, HAW, SAM, SOL; tropical Asia.

S. parviflora (Poir.) Kerguélen, *Lejeunia*, n.s. 120: 161 (1987); *Easter* 81. *Cenchrus parviflorus* Poir., *Encycl.* 6: 52 (1804). *Setaria geniculata* P. Beauv., *Ess. Agrostogr.*: 51 (1812); *Fl. Guam* 225; *List Micro.* 58. *S. gracilis* Kunth in Humb. & Bonpl., *Nov. Gen. Sp.* 1: 109 (1815); *Man. Haw.* 1592.
CRL, EAS, FIJ, HAW, MRN, MRS, NWC, SOL, TON; warm temperate & tropical America.

S. pumila (Poir.) Roem. & Schult., *Syst. Veg.* 2: 891 (1817); *Fl. Pit.* 24; *Fl. Soc.* 344. *Panicum pumilum* Poir., *Encycl.* Suppl. 4: 273 (1818). *P. pallidefuscum* Schumach, *Beskr. Guin. Pl.*: 58 (1827). *Setaria pallidefusca* (Schumach.) Stapf & C. E. Hubb., *Bull. Misc. Inform., Kew* 1930:

259 (1930); *Pl. Tonga* 62; *Fl. Guam* 224; *List Micro.* 58; *Fl. Sol.* 186. *S. glauca sensu Fl. Fiji* 347; *Fl. Soc.* 344. *S. lutescens sensu Fl. N. Cal.* 28.
COO, CRL, FIJ, MCS, MRN, MRS, NWC, PIT, SAM, SCI, SOL, TON, VAN; tropical and warm temperate regions.

S. sphacelata (Schumach.) Stapf & C. E. Hubb. ex M. B. Moss, *Bull. Misc. Inform., Kew* 1929: 195 (1929); *List Micro.* 59; *Fl. Sol.* 186; *Easter* 82; *Fl. Soc.* 344. *Panicum sphacelatum* Schumach., *Beskr. Guin. Pl.*: 78 (1827).
CRL, EAS, FIJ, HAW, MRN, SAM, SCI, SOL; tropical Africa and introduced as a pasture grass.

S. verticillata (L.) P. Beauv., *Ess. Agrostogr.*: 51 (1812); *Fl. N. Cal.* 28; *Fl. Guam* 225; *List Micro.* 59; *Man. Haw.* 1593. *Panicum verticillatum* L., *Sp. Pl.*, ed. 2, 1: 82 (1762). *Chaetochloa verticillata* (L.) Scribn., *Bull. Div. Agrostol. U.S.D.A.* 4: 39 (1897); *Fl. Poly.* 69.
HAW, MRN, MRQ, MRS, NWC, PIT, WAK; tropical & warm temperate regions.

76. Paspalidium

1. Upper glume as long as and covering the upper lemma; spikelets elliptic, 1.7 mm long, distinctly pedicelled, somewhat loose and irregular in 1–2-rowed racemes .... **P. elegantulum**
1. Upper glume shorter than the upper lemma, exposing its tip; spikelets ovate, subsessile, crowded in neatly 2-rowed racemes:
   2. Spikelets 1.8–2.2 mm long; lower floret barren; rhachis extension subulate or bristle-like; leaf-blades 1–4 mm wide .... **P. distans**
   2. Spikelets 2.5–3 mm long; lower floret male; rhachis extension flat, narrowly triangular; leaf-blades 4–8 mm wide .... **P. flavidum**

P. distans (Trin.) Hughes, *Bull. Misc. Inform., Kew* 1923: 317 (1923). *Panicum distans* Trin., *Sp. Gram.* 2: t.172 (1829). *Panicum gracile sensu Fl. N. Cal.* 30.
HAW, NWC; New Guinea, Australia.

P. elegantulum (Mez) Henrard, *Blumea* 3: 435 (1940). *Panicum elegantulum* Mez, *Notizbl. Bot. Gart. Berlin-Dahlem* 7: 59 (1917). *Setaria elegantula* (Mez) Morat, *Adansonia* 22: 270 (2000).
NWC.

P. flavidum (Retz.) A. Camus, Lecomte, *Fl. Indo-Chine* 7: 419 (1922); *List Micro.* 53; *Fl. Sol.* 185. *Panicum flavidum* Retz., *Observ. Bot.* 4: 15 (1791).
CRL, MRN, SOL; tropical Asia.

77. Ixophorus

I. unisetus (J. Presl) Schltdl., *Linnaea* 31: 421 (1862). *Urochloa uniseta* J. Presl in C. Presl, *Reliq. Haenk.* 1: 319 (1830).
HAW; Mexico to Brazil.

78. Dissochondrus

D. biflorus (Hillebr.) Kuntze in Engl. & Prantl, *Nat. Pflanzenfam. Nachtr.* 1: 41 (1897); *Man. Haw.* 1534. *Setaria biflora* Hillebr., *Fl. Hawaiian Isl.*: 503 (1888). *Dissochondrus bifidus* Kuntze, *Revis. Gen. Pl.*: 770 (1891), sphalm.
HAW.

79. Stenotaphrum

1. Inflorescence cylindrical, with racemes of 3–8 spikelets embedded on opposite sides; leaf-blades acute . . . . **S. micranthum**
1. Inflorescence 1-sided; leaf-blades obtuse:
   2. Axis of inflorescence herbaceous, toothed on the back; racemes with 3–8 spikelets . . . . **S. dimidiatum**
   2. Axis of inflorescence corky, entire; racemes commonly reduced to 1 spikelet with rhachis extension . . . . **S. secundatum**

S. dimidiatum (L.) Brongn. in Duperrey, *Voy. Monde Phan.*: 127 (1832); *Fl. N. Cal.* 31; *Pl. Samoa* 30; *Fl. Soc.* 346. *Panicum dimidiatum* L., *Sp. Pl.* 1: 57 (1753).
NWC, SAM, SCI, VAN; shores of the Indian Ocean.

S. micranthum (Desv.) C. E. Hubb. in C. E. Hubb. & R. E. Vaughan, *Grass. Mauritius & Rodriguez*: 73 (1940); *Fl. Niue* 252; *List Micro.* 61; *Fl. Sol.* 186. *Ophiurinella micrantha* Desv., *Opusc. Sci. Phys. Nat.*: 75 (1831). *Stenotaphrum subulatum* Trin., *Mém. Acad. Imp. Sci. St.-Pétersbourg, Sér. 6, Sci. Math., Seconde Pt. Sci. Nat.* 3: 190 (1835); *Fl. N. Cal.* 31; *Pl. Tonga* 57; *Fl. Fiji* 353; *Fl. Soc.* 346.
COO, CRL, FIJ, GIL, HAW, MRN, MRS, NRU, NUE, NWC, SAM, SCI, SOL, TUB, TON, VAN; seashores and littoral woodland from East Africa to the Pacific.

S. secundatum (Walter) Kuntze, *Revis. Gen. Pl.* 2: 794 (1891); *Fl. Raro.* 21; *Fl. N. Cal.* 31; *Pl. Tonga* 57; *Pl. Samoa* 30; *Fl. Fiji* 354; *List Miceo.* 61; *Fl. Sol.* 186; *Man. Haw.* 1598. *Ischaemum secundatum* Walter, *Fl. Carol.*: 249 (1788).
COO, FIJ, HAW, MRN, MRS, NWC, SAM, SCI, SOL, TON, VAN; tropical shores of the Atlantic Ocean.

80. Melinis

1. Spikelets glabrous or occasionally pubescent, with conspicuously ribbed veins; leaves smelling of linseed oil . . . . **M. minutiflora**
1. Spikelets villous with silky pink or silvery hairs, without prominent veins; leaves scentless . . . . **M. repens**

M. minutiflora P. Beauv., *Ess. Agrostogr.*: 54 (1812); *Pl. Tonga* 67; *Fl. Guam* 200; *Fl. Niue* 244; *List Micro.* 49; *Fl. Sol.* 185; *Man. Haw.* 1562; *Easter* 78; *Fl. Soc.* 339.
CRL, EAS, FIJ, HAW, MRN, MRQ, NUE, NWC, SCI, SOL, TON; tropical Africa and widely introduced as a pasture grass.

M. repens (Willd.) Zizka, *Biblioth. Bot.* 138: 55 (1988); *Easter* 79; *Fl. Soc.* 339. *Saccharum repens* Willd., *Sp. Pl.* 1: 332 (1797). *Tricholaena rosea* Nees, *Linnaea* 11, *Litt.-Ber.*: 129 (1837); *List Micro.* 62. *Rhynchelytrum roseum* (Nees) Bews, *World's Grass.*: 223 (1929); *Fl. N. Cal.* 29. *R. repens* (Willd.) C. E. Hubb., *Bull. Misc. Inform., Kew* 1934: 110 (1934); *Pl. Tonga* 62; *Fl. Guam* 222; *Fl. Fann.* 352; *Fl. Fiji* 363; *Fl. Sol.* 185; *Man. Haw.* 1588.
COO, EAS, FIJ, GIL, HAW, LIN, MRN, MRQ, MRS, NRU, NWC, SCI, SOL, TON, TUA; tropical Africa and a common weed throughout the tropics.

81. Digitaria

1. Spikelets borne singly on the raceme rhachis; upper glume less than 1/3 length of spikelet or absent:
  2. Raceme rhachis broadly winged, 2–4 mm wide, the spikelets in pockets on either side of the midrib; sterile lemma with 5 smooth ribbed veins ...... **D. stenotaphrodes**
  2. Raceme rhachis narrowly winged, 0.5–1 mm wide, the spikelets not in pockets:
    3. Racemes paired, 1.5–2.5 cm long; culms prostrate; leaf-blades lanceolate, 1–2 cm long; sterile lemma with 5 smooth veins .... **D. mariannensis**
    3. Racemes 7–16, 6–12 cm long; culms erect; leaf-blades linear, 15–25 cm long; sterile lemma with (5 –)7 scabrid veins .... **D. gaudichaudii**
1. Spikelets borne in groups of 2 or 3:
  4. Raceme rhachis broadly winged, flat, with rounded midrib; spikelets ternate:
    5. Spikelets pubescent, often obscurely so:
      6. Fruit dark brown to black; racemes typically 3 or more .... **D. violascens**
      6. Fruit light brown to light grey; racemes typically paired .... **D. longiflora**
    5. Spikelets entirely glabrous:
      7. Upper glume as long as spikelet, 3–5-veined, equalling the sterile lemma; fruit pallid to light brown .... **D. fuscescens**
      7. Upper glume 2/3 length of spikelet, 3-veined, much shorter than sterile lemma; fruit dark brown to black .... **D. caledonica**
  4. Raceme rhachis with or without narrow wings, the midrib angular; spikelets paired:
    8. Plants clearly perennial (caespitose, rhizomatous or stoloniferous):
      9. Spikelets all quite glabrous, not ribbed; lower glume obvious as a triangular scale up to 1 mm long; plants rhizomatous, mat-forming ..... **D. abyssinica**
      9. Spikelets, or some of them, hairy:
        10. Inflorescence deciduous as a whole, its racemes radiating from a short central axis; racemes naked for the lower 1–9 cm, bearing distant spikelets above .... **D. divaricatissima**
        10. Inflorescence shedding its spikelets at maturity; racemes bearing spikelets almost to the base:
          11. Racemes numerous on a central axis; spikelets villous, the hairs extending 1–2 mm beyond tip; plants caespitose with short rhizomes .... **D. insularis**
          11. Racemes digitate or subdigitate:
            12. Lower spikelet of the pair glabrous, with conspicuously ribbed veins separated by deep slots; upper spikelet long-ciliate; raceme rhachis wingless; plants rhizomatous .... **D. heterantha**

12. Lower spikelet of the pair pubescent to pilose, without conspicuous ribs:
13. Plants densely tufted; culms 35–140 cm high ...... **D. eriantha**
13. Plants robustly stoloniferous, mat-forming; culms 15–30 cm high .... **D. didactyla**
8. Plants annual, sometimes becoming short-lived perennials; spikelets pubescent (often obscurely so) to villous without conspicuous ribs:
14. Raceme rhachis triquetrous without wings, visible between the loosely arranged spikelets:
15. Sterile lemma villous, the hairs extending 1 mm beyond tip; spikelets 2.9–4.2 mm long .... **D. montana**
15. Sterile lemma puberulous with capitate hairs; spikelets 1.5–1.9 mm long .... **D. filiformis**
14. Raceme rhachis narrowly winged, covered by spikelets; sterile lemma pubescent, though sometimes with ciliate fringe or glassy bristles:
16. Rhachis margins smooth; sterile lemma 5-veined, the laterals contiguous (thus apparently 3-veined); upper glume $^1/_3$–$^2/_3$ length of spikelet .... **D. radicosa**
16. Rhachis margins scabrid:
17. Upper glume $^1/_8$–$^1/_4$ length of spikelet; lower glume absent; lower lemma 7-veined; spikelets 2–3 mm long .... **D. setigera**
17. Upper glume $^1/_3$–$^4/_5$ length of spikelet:
18. Lower lemma 5-veined, these equidistant; lower glume 0.5 mm long; upper glume $^1/_2$–$^2/_3$ length of spikelet; spikelets 3.1–3.5 mm long .... **D. mezii**
18. Lower lemma 7-veined:
19. Upper glume $^1/_3$–$^1/_2$ length of spikelet; lower glume tiny, but distinct; spikelets 2–2.5 mm long; inflorescence with a central axis up to 7 cm long .... **D. horizontalis**
19. Upper glume ($^1/_2$–) $^2/_3$–$^4/_5$ length of spikelet; inflorescence digitate or subdigitate:
20. Lower glume absent or obscure; spikelets 1.7–2.5 mm long .... **D. nuda**
20. Lower glume distinct, triangular, 0.2–0.4 mm long; spikelets mostly 2.5–3.5 mm long:
21. Lower lemma of sessile spikelet with ribbed equidistant veins; racemes usually 2, stiff .... **D. bicornis**
21. Lower lemma of sessile spikelet without ribs, the veins usually unequally spaced; racemes 2–12, stiff or flexible .... **D. ciliaris**

D. abyssinica (Hochst. ex A. Rich.) Stapf, *Bull. Misc. Inform., Kew* 1907: 213 (1907).
Panicum abyssinicum Hochst. ex A. Rich., Tent. Fl. Abyss. 2: 360 (1850).
HAW; Africa.

D. bicornis (Lam.) Roem. & Schult., *Syst. Veg.* 2: 470 (1817); *List. Micro.* 37. *Paspalum bicorne* Lam., *Tab. Encycl.* 1: 176 (1791).
CRL, FIJ, HAW, MRN, WAK; tropical Asia, Australia.

D. caledonica Henrard, *Blumea* 1: 100 (1934); *Fl. N. Cal.* 30; *Fl. Fiji* 324.
FIJ, NWC.

D. ciliaris (Retz.) Koeler, *Descr. Gram.*: 27 (1802); *Fl. Fiji* 326; *List Micro.* 37; *Fl. Sol.* 184; *Man. Haw.* 1530; *Easter* 76; *Fl. Soc.* 334. *Panicum ciliare* Retz., *Observ. Bot.* 4: 16 (1786). *P. adscendens* Kunth in Humb. & Bonpl., *Nov. Gen. Sp.* 1: 97 (1815). *Digitaria adscendens* (Kunth) Henrard, *Blumea* 1: 92 (1934); *Fl. Niue* 240; *Fl. Sol.* 184. *D. sanguinalis sensu Pl. Tonga* 56; *Fl. Guam* 208; *Pl. Samoa* 132. *D. henryi sensu List Micro.* 37.
COO, CRL, EAS, FIJ, HAW, MCS, MRN, MRQ, MRS, NRU, NUE, NWC, PHX, SAM, SCI, SOL, TON, TUA, TUB, VAN, WAK, WAL; a common weed throughout the tropics.

D. didactyla Willd., *Enum. Pl.*: 91 (1809); *Fl. Fiji* 323.
FIJ, TON; Madagascar and introduced as a lawn grass.

D. divaricatissima (R. Br.) Hughes, *Bull. Misc. Inform., Kew* 1923: 314 (1923). *Panicum divaricatissimum* R. Br., *Prodr.*: 192 (1810).
HAW; Australia.

D. eriantha Steud., *Flora* 12: 468 (1829). *D. decumbens* Stent, *Bothalia* 3: 150 (1930); *Fl. Fiji* 325. *D. pentzii* Stent, l.c.: 147; *List Micro.* 38.
CRL, FIJ, HAW; South Africa.

D. filiformis (L.) Koeler, *Descr. Gram.*: 26 (1802). *Panicum filiforme* L., *Sp. Pl.* 1: 57 (1753).
TUB; North & South America.

D. fuscescens (J. Presl) Henrard, *Meded. Rijks-Herb.* 61: 8 (1930); *Fl. Fiji* 323; *List Micro.* 37; *Man. Haw.* 1531; *Fl. Soc.* 334. *Paspalum fuscescens* J. Presl in C. Presl, *Reliq. Haenk.* 1: 213 (1830).
COO, CRL, FIJ, HAW, SAM, SCI; tropical Asia.

D. gaudichaudii (Kunth) Henrard, *Meded. Rijks-Herb.* 61: 18 (1930); *Fl. Guam* 209; *List Micro.* 37. *D. stricta* Gaudich., *Voy. Uranie*: 409 (1829), non Roth (1821). *Panicum gaudichaudii* Kunth, *Révis. Gramin.* 2: t.106 (1833). *Digitaria robinsonii* Merr., *Philipp. J. Sci.* 15: 540 (1920); *Fl. Guam* 209. *Syntherisma robinsonii* (Merr.) Hosok., *Trans. Nat. Hist. Soc. Taiwan* 24: 198 (1934). *S. stricta* Hosok., l.c.: 198.
MRN, WAK.

D. heterantha (Hook. f.) Merr., *Lingnaam Agric. Rev.* 1(2): 48 (1923). *Paspalum heteranthum* Hook. f., *Fl. Brit. Ind.* 7: 16 (1896). *Digitaria longissima* Mez, *Repert. Spec. Nov. Regni Veg.* 18: 26 (1922); *List Micro.* 38.
CRL; Southeast Asia, Australia.

D. horizontalis Willd., *Enum. Pl.*: 92 (1809); *Pl. Tonga* 56.
HAW, MRQ, SAM, SCI, TON; West Africa, South America.

D. insularis (L.) Ekman, *Ark. Bot.* 11(4): 17 (1912); *List Micro.* 38; *Fl. Sol.* 184; *Man. Haw.* 1531. *Andropogon insularis* L., *Syst. Nat.*, ed. 10, 2: 1304 (1759). *Trichachne insularis* (L.) Nees, *Agrost. Bras.*: 86 (1829); *Fl. Guam* 226.
CRL, HAW, MRN, MRQ, MRS, SOL, WAK; Central & South America.

D. longiflora (Retz.) Pers., *Syn. Pl.* 1: 85 (1805); *Pl. Samoa* 110; *List Micro.* 38. *Paspalum longiflorum* Retz., *Observ. Bot.* 4: 15 (1786).
CRL, SAM; Old World tropics.

D. mariannensis Merr., *Philipp. J. Sci., Bot.* 9: 54 (1914); *Fl. Guam* 211; *List Micro.* 38. *Syntherisma mariannensis* (Merr.) Hosok., *Trans. Nat. Hist. Soc. Taiwan* 24: 198 (1934).
MRN; Indonesia, Philippines, Australia.

D. mezii Kaneh., *Fl. Micron.*: 401 (1933); *List Micro.* 38. *D. marianensis* Mez, *Bot. Jahrb. Syst.* 59: 1 (1924), non *D. mariannensis* Merr. (1914). *D. latronum* Henrard, *Blumea* 1: 97 (1934). *Syntherisma mezii* (Kaneh.) Hosok., *Trans. Nat. Hist. Soc. Taiwan* 24: 198 (1934).
MRN. No vouchers seen.

D. montana Henrard, *Meded. Rijks-Herb.* 61: 9 (1930); *Fl. N. Cal.* 30. *Panicum collinum* Balansa, *Bull. Soc. Bot. France* 19: 323 (1872). *Digitaria collina* (Balansa) Henrard, *Blumea* 1: 97 (1934), non Salisb. (1796).
NWC.

D. nuda Schumach., *Beskr. Guin. Pl.*: 45 (1827).
CRL, NWC; Africa.

D. radicosa (J. Presl) Miq., *Fl. Ned. Ind.* 3: 437 (1857); *Fl. Fiji* 325; *List Micro.* 38; *Fl. Soc.* 334. *Panicum radicosum* J. Presl in C. Presl, *Reliq. Haenk.* 1: 297 (1830). *Digitaria propinqua* Gaudich., *Voy. Uranie*: 410 (1829), non (R. Br.) P. Beauv. (1812); *Fl. N. Cal.* 30. *Panicum timorense* Kunth, *Enum. Pl.* 1: 83 (1833). *Digitaria timorensis* (Kunth) Balansa, *J. Bot. (Morot)* 4: 138 (1890); *Fl. Raro.* 18; *Fl. Guam* 211; *Fl. Sol.* 184.
COO, CRL, FIJ, GIL, HAW, MRN, MRQ, MRS, NRU, NWC, SAM, SCI, SOL, TON, VAN; tropical Asia.

D. setigera Roem. & Schult., *Syst. Veg.* 2: 474 (1817); *Fl. Fiji* 326; *Fl. Pit.* 23; *List Micro.* 39; *Fl. Sol.* 184; *Man. Haw.* 1531; *Easter* 76; *Fl. Soc.* 334. *Panicum pruriens* Fisch. ex Trin., *Gram. Panic.*: 77 (1826). *P. microbachne* J. Presl in C. Presl, *Reliq. Haenk.* 1: 298 (1830). *Digitaria pruriens* (Fisch. ex Trin.) Büse in Miq., *Pl. Jungh.*: 379 (1854); *Fl. Raro.* 18; *Fl. N. Cal.* 30; *Pl. Tonga* 56; *Fl. Guam* 209; *Fl. Niue* 240; *Fl. Sol.* 184. *Syntherisma pruriens* (Fisch. ex Trin.) Arthur, *Torreya* 19: 83 (1919); *Fl. Poly.* 72. *Digitaria microbachne* (J. Presl) Henrard *Meded. Rijks-Herb.* 61: 13 (1930); *Fl. Raro.* 18; *Pl. Samoa* 132; *Fl. Sol.* 184. *D. pruriens* var. *microbachne* (J. Presl) Fosberg, *Phytologia* 5: 289 (1955); *Fl. Guam* 210. *D. microbachne* subsp. *calliblepharata* Henrard, *Monogr. Digitaria*: 452 (1950). *D. setigera* var.

calliblepharata (Henrard) Veldkamp, *Blumea* 21(1): 40 (1973); *List Micro*. 39.
COO, CRL, EAS, FIJ, GIL, HAW, MCS, MRN, MRQ, MRS, NRU, NUE, NWC, PHX, PIT, SAM, SCI, SOL, TOK, TON, TUA, TUB, VAN, WAK; tropical Asia.

D. stenotaphrodes (Steud.) Stapf, *Bull. Misc. Inform., Kew* 1906: 77 (1906). *Panicum stenotaphrodes* Steud., *Syn. Pl. Glumac.* 1: 41 (1853). *Digitaria pacifica* Stapf, *Bull. Misc. Inform., Kew* 1906: 77 (1906); *List Micro*. 38. *Syntherisma stenotaphrodes* (Steud.) Chase, *Proc. Biol. Soc. Wash*. 19: 191 (1906). *S. pelagica* F. Br., *Bernice P. Bishop Mus. Bull*. 84: 73 (1931); *Fl. Poly*. 72.
COO, CRL, GIL, HBI, LIN, PHX, SCI, TUA.

D. violascens Link, *Hort. Berol*. 1: 229 (1827); *Fl. Guam* 211; *Fl. Niue* 241; *Fl. Fiji* 324; *List Micro*. 40; *Man. Haw*. 1532; *Easter* 77; *Fl. Soc*. 335. *Syntherisma helleri* Nash, *Minnesota Bot. Stud*. 1: 798 (1897).
CRL, EAS, FIJ, HAW, MRN, MRQ, NRU, NUE, NWC, SAM, SCI; tropical Asia, America.

82. Pennisetum

1. Inflorescence reduced to a cluster of 2–4 subsessile spikelets enclosed in uppermost leaf-sheath, with long protruding filaments and stigmas .... **P. clandestinum**
1. Inflorescence a spiciform panicle, conspicuously exserted:
  2. Involucral clusters persistent .... **P. glaucum**
  2. Involucral clusters readily deciduous:
    3. Rhachis of panicle with decurrent wings below scars of fallen involucres; upper lemma coriaceous, shiny, readily deciduous .... **P. polystachion**
    3. Rhachis cylindrical or with angular ribs, but these not expanded into winged brackets below the scars; upper lemma firmly membranous, dull, resembling the lower:
      4. Panicle oblong; spikelets 9–14 mm long; involucral bristles 40–70 mm long, plumose .... **P. villosum**
      4. Panicle linear; spikelets 4–8 mm long; involucral bristles 6–40 mm long:
        5. Involucres borne on a linear stipe 1–3 mm long:
          6. Leaf-blades convolute, 1–3 mm wide, the midrib noticeably thickened on upper surface .... **P. setaceum**
          6. Leaf-blades flat, 4–12 mm wide, the midrib not conspicuously wide .... **P. orientale**
        5. Involucres without a stipe, at most with a conical or oblong foot 0.5(– 1) mm long:
          7. Peduncle pubescent to hirsute below the panicle; involucres containing 1–5 spikelets, the bristles glabrous or sparsely ciliate, 10–40 mm long; leaf-blades 20–40 mm wide .... **P. purpureum**
          7. Peduncle glabrous below the panicle; involucres containing 1 spikelet, the bristles glabrous:
            8. Involucres with 1 bristle noticeably longer than the rest; culms 75–200 cm high; leaf-blades 1–5 mm wide .... **P. complanatum**
            8. Involucres with the inner bristles subequal:

9. Ligule prominent, ciliate; leaf-blades 2–5 mm wide; culms 15–30 cm high .... **P. articulare**
9. Ligule absent or represented by a puberulous rim; leaf-blades 7–30 mm wide:
    10. Culms erect, 200–300 cm high; panicles terminal and axillary .... **P. macrostachyum**
    10. Culms decumbent, 30–200 cm long; panicle terminal .... **P. henryanum**

P. articulare Trin. in Spreng., *Neue Entd.* 2: 77 (1821); *Fl. Poly.* 63. *P. identicum* Steud. ex Jard., *Mém. Soc. Sci. Nat. Cherbourg* 5: 299 (1857), nom. nud. *P. simeonis* F. Br., *Bernice P. Bishop Mus. Bull.* 84: 61 (1931). *P. simeonis* var. *intermedium* F. Br., l.c.: 61. *P. simeonis* var. *pedicellatum* F. Br., l.c.: 61. *P. simeonis* var. *purpureum* F. Br., l.c.: 62.
MRQ.

P. clandestinum Hochst. ex Chiov., *Annuario Reale Ist. Bot. Roma* 8: 41 (1903); *Man. Haw.* 1578; *Easter* 81.
EAS, HAW; eastern Africa and widely introduced.

P. complanatum (Nees) Hemsl., *Biol. Cent.-Amer., Bot.* 3: 507 (1885). *Gymnothrix complanata* Nees, *Bonplandia* 3: 83 (1855).
HAW; Central America.

P. glaucum (L.) R. Br., *Prodr.*: 195 (1810). *Panicum glaucum* L., *Sp. Pl.* 1: 56 (1753). *P. americanum* L., l.c.: 56. *Setaria glauca* (L.) P. Beauv., *Ess. Agrostogr.*: 51 (1812). *Pennisetum americanum* (L.) Leeke, *Z. Naturwiss.* 79:52 (1907); *Fl. Fiji* 359; *List Micro.* 55.
FIJ, HAW, MRN; Africa. Bulrush millet, cultivated as a cereal in dry regions.

P. henryanum F. Br., *Bernice P. Bishop Mus. Bull.* 84: 61 (1931). *P. henryanum* var. *longisetum* F. Br., l.c.: 61. *P. henryanum* var. *pluristylum* F. Br., l.c.: 61.
MRQ.

P. macrostachyum (Brongn.) Trin., *Mém. Acad. Imp. Sci. St.-Pétersbourg, Sér. 6, Sci. Math., Seconde Pt. Sci. Nat.* 3(2): 177 (1835); *Fl. Niue* 250; *Fl. Sol.* 185. *Gymnothrix macrostachys* Brongn., Duperrey, *Voy. Monde Phan.*: 104 (1831).
FIJ, HAW, NUE, SOL; Malesia.

P. orientale Rich. in Pers., *Syn. Pl.* 1: 72 (1805). *P. triflorum* Steud., *Syn. Pl. Glumac.* 1: 107 (1854); *List Micro.* 55.
MRN; Middle East, India.

P. polystachion (L.) Schult., *Mant.* 2: 146 (1824); *Fl. Guam* 221; *Fl. Fiji* 359; *List Micro.* 55; *Fl. Sol.* 185; *Man. Haw.* 1579. *Panicum polystachion* L., *Syst. Nat.*, ed. 10, 2: 870 (1759). *Cenchrus setosus* Sw., *Prodr.*: 26 (1788). *Pennisetum setosum* (Sw.) Rich. in Pers., *Syn. Pl.* 1: 72 (1805); *Fl. Guam* 220; *Fl. Niue* 250; *Fl. Sol.* 185. *P. polystachion* f. *viviparum* Fosberg & Sachet, *Micronesica* 18: 86 (1984).
CRL, FIJ, GIL, HAW, MCS, MRN, MRS, NUE, SOL, WAK; throughout the tropics.

P. purpureum Schumach., *Beskr. Guin. Pl.*: 55 (1827); *Fl. N. Cal.* 31; *Fl. Guam* 220; *Fl. Niue* 250; *Fl. Fiji* 358; *List Micro.* 55; *Man. Haw.* 1579; *Fl. Soc.* 342.
CRL, FIJ, GIL, HAW, MRN, MRS, NUE, NWC, SAM, SCI, SOL; Africa and widely introduced as a forage plant.

P. setaceum (Forssk.) Chiov., *Boll. Soc. Bot. Ital.* 1923: 113 (1923); *Fl. Fiji* 359; *Man. Haw.* 1581. *Phalaris setacea* Forssk., *Fl. Aegypt.-Arab.*: 17 (1775).
FIJ, HAW; Africa, Middle East.

P. villosum R. Br. ex Fresen., *Mus. Senckenberg.* 2: 134 (1837).
HAW; Northeast Africa, Arabia, introduced elsewhere as an ornamental.

83. Cenchrus

1. Bristles or spines antrorsely barbed:
   2. Inner bristles ciliate, flexuous, united only at the base to form a shallow disc 0.5–1.5 mm in diameter (often mistaken for *Pennisetum* if disc and basal flattening of inner bristles overlooked) . . . . **C. ciliaris**
   2. Inner bristles glabrous, rigid, flattened, connate for $^{1}/_{4}$–$^{2}/_{3}$ of their length to form a cup . . . . **C. setigerus**
1. Bristles or spines retrorsely barbed:
   3. Spines connate for at least $^{1}/_{2}$ their length to form a globose bur:
      4. Bur with spines emerging irregularly from its surface . . . . **C. tribuloides**
      4. Bur subtended by a whorl of finer spines at its base (and sometimes a few on surface in *C. echinatus*):
         5. Outer bristles subequal to inner; burs crowded in the inflorescence . . **C. brownii**
         5. Outer bristles about $^{1}/_{2}$ as long as inner; burs interrupted in the inflorescence . . . . **C. echinatus**
   3. Spines connate only at the base to form a shallow cup:
      6. Burs ovate to globose, with 8–20 narrow subterete inner spines; foot of bur oblong . . . . **C. caliculatus**
      6. Burs fusiform to turbinate, with 6–10 broad flat inner spines; foot of bur broadly conical . . . . **C. agrimonioides**

C. agrimonioides Trin., *Gram. Panic.*: 72 (1826); *Man. Haw.* 1511. *C. fusiformis* Nees & Meyen, *Gramineae*: 38 (1841), nom. superfl. *C. caliculatus* var. *uniflorus* Hillebr., *Fl. Hawaiian. Isl.*: 505 (1888). *C. agrimonioides* var. *laysanensis* F. Br., *Bernice P. Bishop Mus. Bull.* 81: 20 (1931). *C. pedunculatus* O. Deg. & Whitney in O. Deg., *Fl. Hawaii., Fam.*: 47 (1936). *C. laysanensis* (F. Br.) H. St.John, *Phytologia* 31: 22 (1975).
HAW.

C. brownii Roem. & Schult., *Syst. Veg.* 2: 258 (1817); *Fl. Guam* 208; *List Micro.* 33; *Fl. Sol.* 184.
CRL, MRN, MRS, SOL, WAK; Southeast Asia, tropical America.

C. caliculatus Cav., *Icon.* 5: 39, t. 463 (1799); *Fl. Raro.* 17; *Fl. N. Cal.* 31; *Pl. Tonga* 63; *Fl. Niue* 236; *Fl. Fiji* 355; *Fl. Pit.* 22; *Fl. Soc.* 330. *C. anomoplexis* Labill., *Sert. Austro-Caledon.*: 14, t. 19 (1824), non Desf. (1799). *Pennisetum caliculatum* (Cav.) Spreng., *Syst. Veg.* 1: 303 (1825). *Cenchrus laniflorus* Steud., *Syn. Pl. Glumac.* 1: 110 (1854). *C. taitensis* Steud., l.c.: 419. *C. australis* var. *latifolius* Drake, *Fl. Polynésie Franç.*: 252 (1893). *Pennisetum marquisense* F. Br., *Bernice P. Bishop Mus. Bull.* 84: 63 (1931).
COO, FIJ, HAW, MRQ, NUE, NWC, PIT, SAM, SCI, TON, TUB, VAN; New Guinea, Australia.

C. ciliaris L., *Mant. Pl.* 2: 302 (1771); *Fl. Niue* 231; *Fl. Fiji* 357; *Fl. Sol.* 184; *Man. Haw.* 1511.
FIJ, HAW, MRQ, NUE, NWC, PHX, SCI, SOL; Africa to India and widely introduced as a forage plant.

C. echinatus L., *Sp. Pl.* 2: 1050 (1753); *Fl. Poly.* 65; *Fl. Raro.* 17; *Pl. Tonga* 63; *Fl. Guam* 205; *Fl. Niue* 236; *Pl. Samoa* 147; *Fl. Fann.* 357; *Fl. Fiji* 357; *List Micro.* 33; *Fl. Sol.* 184; *Man. Haw.* 1512; *Easter* 74; *Fl. Soc.* 330. *C. hillebrandianus* Hitchc., *Mem. Bernice Puauhi Bishop Mus.* 84: 66 (1931). *C. echinatus* var. *glabratus* F. Br., *Bernice P. Bishop Mus. Bull.* 84: 66 (1931). *C. echinatus* var. *hillebrandianus* (Hitchc.) F. Br., l.c.: 65. *C. echinatus* var. *pennisetoides* F. Br., l.c.: 66.
COO, CRL, EAS, FIJ, GIL, HAW, LIN, MCS, MRN, MRQ, MRS, NRU, NUE, NWC, PHX, SAM, SCI, SOL, TON, TUA, TUB, VAN, WAK; South America and widespread as a weed.

C. setigerus Vahl, *Enum. Pl.* 2: 395 (1806).
HAW; eastern Africa, Middle East.

C. tribuloides L., *Sp. Pl.* 2: 1050 (1753).
HAW, TON; eastern USA, Caribbean.

84. Anthephora

A. hermaphrodita (L.) Kuntze, *Revis. Gen.* Pl. 2: 759 (1891). *Tripsacum hermaphroditum* L., *Syst. Nat.*, ed. 10, 2: 1261 (1759).
HAW; Florida to Brazil.

85. Spinifex

S. sericeus R. Br., *Prodr.*: 198 (1810). *S. hirsutus sensu Fl. N. Cal.* 28. *S. littoreus sensu Pl. Tonga* 64.
NWC, TON; Australia, New Zealand.

ISACHNEAE

86. Isachne

1. Florets dissimilar in texture, the lower herbaceous to cartilaginous and usually longer, the upper crustaceous:

2. Glumes pilose at the tip; upper lemma pilose .... **I. comata**
2. Glumes and lemmas glabrous or puberulous:
   3. Culms glandular below the nodes; spikelets 1.5–1.6 mm long, obovoid .... **I. pulchella**
   3. Culms without glands:
      4. Culm nodes glabrous; spikelets 1.75–2.7 mm long, globose or ellipsoid; leaf-blades 7–9-veined .... **I. globosa**
      4. Culm nodes pubescent; spikelets 1.3–2 mm long, obovate; leaf-blades 5-veined .... **I. minutula**
1. Florets similar, both crustaceous:
   5. Leaf-blades coriaceous, stiff, with thick cartilaginous and often crenate margins; lemmas pubescent on base and margins; culms robust, rambling .. **I. distichophylla**
   5. Leaf-blades herbaceous, the margins not or inconspicuously cartilaginous:
      6. Spikelets globose, 1–1.4 mm long; culms slender, prostrate; leaf-blades lanceolate, 1.2–2.5 cm long; glumes hispidulous .... **I. confusa**
      6. Spikelets elliptic to ovate; leaf-blades linear to lanceolate, 1.5–20 cm long:
         7. Lemmas hairy all over; culms decumbent:
            8. Lemmas pilose; spikelets 2 mm long; leaf-blades 2–4 mm wide .. **I. pallens**
            8. Lemmas puberulous:
               9. Spikelets 1.2–1.5 mm long; panicle open, its branches ascending, the spikelets spreading .... **I. villosa**
               9. Spikelets 2–3.5 mm long; panicle somewhat contracted, its branches erect or ascending, the spikelets appressed .... **I. kunthiana**
         7. Lemmas glabrous on the back with pubescent margins:
            10. Culms robust, ascending; leaf-blades 6–20 mm wide; spikelets 1–1.3 mm long .... **I. vitiensis**
            10. Culms slender, creeping; leaf-blades 5–8 mm wide; spikelets 2 mm long .... **I. carolinensis**

I. carolinensis Ohwi, *Bot. Mag. (Tokyo)* 55: 540 (1941); *List Micro*. 45.
CRL, SOL.

I. comata Hack., *Hooker's Icon. Pl.* 19: t. 1866 (1889).
NWC, VAN.

I. confusa Ohwi, *Bull. Tokyo Sci. Mus.* 18: 14 (1947); *List Micro.* 45. *I. purpurascens* Glassman, *Bernice P. Bishop Mus. Bull.* 209: 130 (1952). *I. confusa* var. *purpurascens* (Glassman) Fosberg & Sachet, *Micronesica* 18: 53 (1984); *List Micro*. 45.
CRL; Southeast Asia, Australia.

I. distichophylla Hillebr., *Fl. Hawaiian Isl.*: 504 (1888); *Fl. Raro.* 19; *Man. Haw.* 1555. *Panicum wiliwilinuiense* H. St.John, *Phytologia* 63: 372 (1987).
COO, HAW.

I. globosa (Thunb.) Kuntze, *Revis. Gen. Pl.* 2: 778 (1891); *Fl. Fiji* 364; *List Micro.* 45; *Fl. Sol.* 184. *Milium globosum* Thunb. ex Murray, *Syst. Veg.*, ed. 14: 109 (1784). *Isachne miliacea* Roth in Roem. & Schult., *Syst. Veg.* 2: 476 (1817); *Fl. Guam* 213. *I. ponapensis* Hosok.,

Trans. Nat. Hist. Soc. Taiwan 24: 200 (1934); *List Micro.* 45. *I. globosa* var. *ciliaris* Ohwi, *Bot. Mag. (Tokyo)* 55: 540 (1941); *List Micro.* 45.
COO, CRL, FIJ, MRN, NWC, SOL; tropical Asia, Australia, New Zealand.

I. kunthiana (Steud.) Miq., *Fl. Ned. Ind.* 3: 460 (1857); *Fl. Sol.* 184. *Panicum kunthianum* Steud., *Syn. Pl. Glumac.* 1: 96 (1854).
SOL; tropical Asia.

I. minutula (Gaudich.) Kunth, *Révis. Gramin.* 2: 117 (1831). *Panicum minutulum* Gaudich., *Voy. Uranie*: 410 (1830). *I. miliacea* var. *minutula* (Gaudich.) Fosberg & Sachet, *Micronesica* 18: 55 (1984); *List Micro.* 45.
CRL, MRN; tropical Asia.

I. pallens Hillebr., *Fl. Hawaiian Isl.*: 504 (1888); *Man. Haw.* 1555.
HAW.

I. pulchella Roth in Roem. & Schult., *Syst. Veg.* 2: 476 (1817); *Fl. Guam* 213. *Panicum pulchellum* (Roth) Spreng., *Syst. Veg.* 1: 322 (1825), non Raddi (1823). *Isachne dispar* Trin., *Sp. Gram.* 1(8): t. 86 (1828).
MRN; tropical Asia.

I. villosa (Hitchc.) Reeder, *J. Arnold Arbor.* 29: 314 (1948). *I. brassii* var. *villosa* Hitchc., *Brittonia* 2: 123 (1936).
SOL; New Guinea.

I. vitiensis Rendle, *J. Linn. Soc., Bot.* 39: 181 (1909); *Fl. Fiji* 364.
FIJ, VAN.

Note. The species tend to intergrade, but the genus has never been fully revised and its taxonomy is still unclear.

ERIACHNEAE

87. Eriachne

E. pallescens R. Br., *Prodr.*: 184 (1810); *List Micro.* 43.
CRL; tropical Asia, Australia.

ARUNDINELLEAE

88. Garnotia

1. Panicle open, ovate, with stiffly spreading branches .... **G. divergens**
1. Panicle contracted, linear to lanceolate, with erect or ascending branches:

2. Lemma awn terminating in a thread-like crinkled portion:
   3. Ligule 1–1.5 mm long .... **G. cheesemanii**
   3. Ligule 0.3–0.8 mm long:
      4. Panicle branches borne singly on the axis .... **G. linearis**
      4. Panicle branches 2 or more at each node of the axis .... **G. stricta**
2. Lemma with or without a uniformly tapering awn:
   5. Leaf-blades 1–3 mm wide, 2–6 cm long:
      6. Leaf-blades involute or folded, not spreading at maturity; lemma with or without an awn up to 5 mm long .... **G. depressa**
      6. Leaf-blades flat, widely spreading or reflexed at maturity; lemma with an awn 4–9 mm long .... **G. gracilis**
   5. Leaf-blades, at least some of them, more than 4 mm wide and 6 cm long:
      7. Leaf-sheaths densely villous above; lemmas usually awnless .... **G. villosa**
      7. Leaf-sheaths glabrous or sparsely hairy:
         8. Lemma awnless or with an awn up to 2 mm long:
            9. Leaf-blades mostly 4–9 cm long, glabrous on collar .... **G. st-johnii**
            9. Leaf-blades mostly 15–30 cm long, usually hispid on collar ... **G. stricta**
         8. Lemma with an awn 4–15 mm long:
            10. Leaf-blades flat, at least some 8–12 mm wide; culms mostly 60–90 cm high .... **G. foliosa**
            10. Leaf-blades flat or folded, 2–6(– 8) mm wide:
               11. Plants mat-forming or densely tufted; culms 15–30 cm high, often trailing .... **G. raiateensis**
               11. Plants loose; culms 30–70 cm high, erect or geniculately ascending .... **G. acutigluma**

G. acutigluma (Steud.) Ohwi, *Bot. Mag. (Tokyo)* 55: 393 (1941); *Man. Haw.* 1548. *Urachne acutigluma* Steud., *Syn. Pl. Glumac.* 1: 121 (1854). *Garnotia sandwicensis* Hillebr., *Fl. Hawaiian Isl.*: 513 (1888).
HAW; Southeast Asia.

G. cheesemanii Hack., *Trans. Linn. Soc. London, Bot., Ser.* 2, 6: 303 (1903). *G. rarotongensis* Whitney, *Occas. Pap. Bernice Puauhi Bishop Mus.* 13: 77 (1937). *G. cheesemanii* var. *rarotongensis* (Whitney) Santos, *Nat. Appl. Sci. Bull. Univ. Philipp.* 10: 76 (1950).
COO.

G. depressa J. W. Moore, *Bernice P. Bishop Mus. Bull.* 102: 18 (1933). *G. mucronata* Swallen, *J. Wash. Acad. Sci.* 26: 178 (1936).
FIJ, SCI.

G. divergens Swallen, *J. Arnold. Arbor.* 31: 143 (1950); *Fl. Fiji* 316.
FIJ.

G. foliosa Swallen, *J. Arnold Arbor.* 31: 142 (1950); *Fl. Fiji* 318.
FIJ.

G. gracilis Swallen, *J. Arnold Arbor.* 31: 142 (1950); *Fl. Fiji* 316.
FIJ.

G. linearis Swallen, *J. Arnold Arbor.* 31: 143 (1950); *Fl. Fiji* 318. *G. munroana* Santos, *Nat. Appl. Sci. Bull. Univ. Philipp.* 10: 64 (1950). *G. solitaria* Santos, l.c.: 95.
FIJ.

G. raiateensis J. W. Moore, *Bernice P. Bishop Mus. Bull.* 102: 18 (1933); *Fl. Soc.* 337.
SCI.

G. st-johnii J. W. Moore, *Bernice P. Bishop Mus. Bull.* 226: 1 (1963).
SCI.

G. stricta Brongn. in Duperrey, *Voy. Monde Phan.* 113, t. 21 (1832); *Fl. Guam* 211; *List Micro.* 43; *Fl. Soc.* 338. *G. stricta* var. *marianarum* Santos, *Nat. Appl. Sci. Bull. Univ. Philipp.* 10: 53 (1950).
CRL, FIJ, HAW, MRN, SCI; tropical Asia.

G. villosa Swallen, *J. Arnold Arbor.* 31: 143 (1950); *Fl. Fiji* 318.
FIJ.

ANDROPOGONEAE

89. Saccharum

1. Upper lemma with an awn 0.5–4 mm long .... **S. maximum**
1. Upper lemma awnless:
   2. Axis of inflorescence glabrous to puberulous .... **S. officinarum**
   2. Axis of inflorescence hirsute with few to many long hairs .... **S. spontaneum**

S. maximum (Brongn.) Trin., *Mém. Acad. Imp. Sci. St.-Pétersbourg, Sér. 6, Sci. Math., Seconde Pt. Sci. Nat.* 4: 92 (1836); *Fl. Soc.* 342. *Erianthus maximus* Brongn. in Duperrey, *Voy. Monde Phan.*: 97 (1831); *Fl. Poly.* 58; *Fl. N. Cal.* 25; *Pl. Samoa* 33; *Fl. Fiji* 367. *Saccharum pedicellare* Trin., *Mém. Acad. Imp. Sci. St.-Pétersbourg, Sér. 6, Sci. Math.* 3: 310 (1833); *Fl. Poly.* 57. *Erianthus maximus* var. *seemannii* Hack. in A. DC. & C. DC., *Monogr. Phan.* 6: 139 (1889); *Fl. Poly.* 58. *E. pedicellaris* (Trin.) Hack., l.c.: 137; *Fl. Poly.* 57. *E. pedicellaris* var. *rapensis* F. Br., *Bernice P. Bishop Mus. Bull.* 84: 57 (1931); *Fl. Poly.* 57.
COO, FIJ, MRQ, NWC, SAM, SCI, TUB, VAN.

S. officinarum L., *Sp. Pl.* 1: 54 (1753); *Fl. Poly.* 58; *Fl. Raro.* 21; *Fl. N. Cal.* 25; *Pl. Tonga* 65; *Fl. Guam* 238; *Pl. Samoa* 131; *Fl. Fiji* 369; *Fl. Pit.* 24; *List Micro.* 57; *Fl. Sol.* 185; *Easter* 81; *Fl. Soc.* 343.
COO, CRL, EAS, FIJ, GIL, HAW, MRN, MRQ, MRS, NUE, NWC, PIT, SAM, SCI, SOL, TON, TUB, VAM; cultivated sugar cane of the tropics.

S. spontaneum L., *Mant. Pl.* 2: 183 (1771); *Fl. N. Cal.* 25; *Fl. Guam* 237; *List Micro.* 57; *Fl. Sol.* 185; *Man. Haw.* 1589; *Fl. Soc.* 343. *S. insulare* Brongn. in Duperrey, *Voy. Monde Phan.*: 99 (1831). *S. edule* Hassk., *Flora* 25 (Beibl. 2): 3 (1842); *Pl. Samoa* 33; *Fl. Fiji* 370; *Fl. Sol.* 185. *S. spontaneum* var. *edule* (Hassk.) K. Schum. & Lauterb., *Fl. Schutzgeb. Südsee*: 166 (1901). *S. robustum* Grassl, *J. Arnold Arbor.* 27: 234 (1946); *Fl. Sol.* 185. *S. spontaneum* var. *insulare* (Brongn.) Fosberg & Sachet, *Micronesica* 18: 89 (1984); *List Micro.* 57.
CRL, FIJ, HAW, MRN, NWC, SAM, SCI, SOL, VAN; Old World tropics. *Saccharum edule* has an abortive inflorescence, which is cooked and eaten as a vegetable.

90. Miscanthus

1. Inflorescence axis more than $^{2}/_{3}$ length of panicle; spikelets 2.5–4(– 6) mm long .... **M. floridulus**
1. Inflorescence axis up to $^{1}/_{2}$ length of panicle; spikelets 4–6.5 mm long ..... **M. sinensis**

M. floridulus (Labill.) K. Schum. & Lauterb., *Fl. Schutzgeb. Südsee*: 166 (1901); *Pl. Tonga* 64; *Fl. Guam* 236; *Fl. Niue* 246; *Pl. Samoa* 134; *Fl. Fiji* 366; *List Micro.* 50; *Fl. Sol.* 185; *Fl. Soc.* 340. *Saccharum floridulum* Labill., *Sert. Austro-Caledon.*: 13, t.18 (1824). *Miscanthus japonicus* Andersson, *Öfvers. Förh. Kongl. Svenska Vetensk.-Akad.* 12: 166 (1855); *Fl. Poly.* 55; *Fl. Raro.* 19; *Fl. N. Cal.* 25.
COO, CRL, FIJ, GIL, MRN, MRQ, MRS, NUE, NWC, SAM, SCI, SOL, TON, TUA, TUB, VAM; Southeast Asia.

M. sinensis Andersson, *Öfvers. Förh. Kongl. Svenska Vetensk.-Akad.* 12: 166 (1855); *List Micro.* 50.
MRS; Southeast Asia.

91. Imperata

1. Panicum spiciform, linear; branches up to 2 cm long, appressed and mostly concealed by spikelets; stamens 2 .... **I. cylindrica**
1. Panicle narrowly lanceolate; branches up to 6 cm long, ascending, distinct; stamen 1. ... . **I. conferta**

I. conferta (J. Presl) Ohwi, *Bot. Mag. (Tokyo)* 55: 549 (1941); *Fl. Fiji* 365; *List Micro.* 44. *Saccharum confertum* J. Presl in C. Presl, *Reliq. Haenk.* 1: 346 (1830). *Imperata exaltata sensu Fl. N. Cal.* 25.
CRL, FIJ, MRN, NWC, SAM, SOL, TON, VAN; Southeast Asia.

I. cylindrica (L.) Raeusch., *Nomencl. Bot.*, ed. 3: 10 (1797): *Fl. N. Cal.* 25; *Pl. Tonga* 64; *List Micro.* 45; *Fl. Sol.* 184. *Lagurus cylindricus* L., *Syst. Nat.*, ed. 10, 2: 878 (1759).
FIJ, MRN, NWC, SAM, SOL, TON, VAN, a notorious weed from the Old World tropics.

92. Eulalia

E. aurea (Bory) Kunth, *Révis. Gramin.* 1: 160 (1829). *Andropogon aureus* Bory, *Voy. Îles Afrique* 1: 367 (1804). *Saccharum fulvum* R. Br., *Prodr.*: 203 (1810). *Eulalia fulva* (R. Br.) Kuntze, *Revis. Gen. Pl.*: 775 (1891); *List Micro.* 43.
MRN; Africa, Vietnam, Philippines, Australia.

93. Polytrias

P. indica (Houtt.) Veldkamp, *Blumea* 36: 180 (1991). *Phleum indicum* Houtt., *Nat. Hist.* 13: 198 (1782). *Andropogon amaurus* Büse in Miq., *Pl. Jungh.* 3: 360 (1854). *Polytrias amaura* (Büse) Kuntze, *Revis. Gen. Pl.*: 788 (1891); *Fl. Guam* 237; *Fl. Fiji* 368; *List Micro.* 56.
CRL, FIJ, MRN; Southeast Asia.

94. Pogonatherum

1. Spikelets 1.3–2 mm long; callus hairs about 2 mm long; stamen 1. . . . . **P. crinitum**
1. Spikelets 2.5–3 mm long; callus hairs up to 1.5 mm long; stamens 2 . . . . . . **P. paniceum**

P. crinitum (Thunb.) Kunth, *Enum. Pl.* 1: 478 (1833); *List Micro.* 56; *Fl. Sol.* 185. *Andropogon crinitus* Thunb., *Fl. Jap.*: 40 (1784).
MRN, SOL; Southeast Asia.

P. paniceum (Lam.) Hack., *Allg. Bot. Z. Syst.* 12: 178 (1906); *Fl. Sol.* 185. *Saccharum paniceum* Lam., *Encycl.* 1: 595 (1785).
SOL; tropical Asia.

95. Microstegium

1. Culms robust, 100–170 cm long; leaf-blades 10–17 mm wide; inflorescence of 10–40 racemes; internodes, pedicels and callus ciliate; spikelets 3–4 mm long; upper lemma with an awn 4–6 mm long . . . . **M. spectabile**
1. Culms slender, 10–14 cm long; leaf-blades 2–10 mm wide; inflorescence of 2–11 racemes; spikelets 2–3 mm long; upper lemma with an awn 8–15 mm long:
   2. Internodes, pedicels and callus glabrous; upper glume mucronate . . . . . **M. glabratum**
   2. Internodes, pedicels and callus ciliate; upper glume with an awn 1–3 mm long . . . . **M. tenue**

M. glabratum (Brongn.) A. Camus, *Ann. Soc. Linn. Lyon, n.s.* 68: 201 (1922); *Fl. N. Cal.* 25; *Fl. Guam* 236; *Fl. Fiji* 371; *List Micro.* 49; *Fl. Soc.* 339. *Eulalia glabrata* Brongn. in Duperrey, *Voy. Monde Phan.*: 93 (1831). *Nemastachys taitensis* Steud., *Syn. Pl. Glumac.* 1: 357 (1854). *Pollinia gracillima* Hack. in A. DC & C. DC., *Monogr. Phan.* 6: 179 (1889). *Eulalia gracillima* (Hack.) Kuntze, *Revis. Gen. Pl.* 2: 775 (1831). *Microstegium gracillimum* (Hack.) A. Camus, *Ann. Soc. Linn. Lyon, n.s.* 68: 201 (1922).
CRL, FIJ, MRN, NWC, SAM, SCI; Philippines.

M. spectabile (Trin.) A.Camus, *Ann. Soc. Linn. Lyon, n.s.* 68: 200 (1922); *Fl. Sol.* 185. *Pollinia spectabilis* Trin., *Mém. Acad. Imp. Sci. St.-Pétersbourg, Sér. 6, Sci. Math.* 2: 305 (1833). *Eulalia spectabilis* (Trin.) Kuntze, *Revis. Gen. Pl.* 2: 775 (1891).
CRL, SOL; New Guinea.

M. tenue (Trin.) Hosok., *Trans. Nat. Hist. Soc. Taiwan* 28: 150 (1938); *List Micro.* 49. *Pollinia tenuis* Trin., *Mém. Acad. Imp. Sci. St.-Pétersbourg, Sér. 6, Sci. Math.* 2: 306 (1833).
CRL; Philippines.

96. Sorghum

1. Culm nodes bearded .... **S. nitidum**
1. Culm nodes glabrous or pubescent:
   2. Racemes tough or tardily disarticulating:
      3. Grain large, commonly exposed by the gaping glumes; sessile spikelet persistent .. .. **S. bicolor**
      3. Grain enclosed by the glumes; sessile spikelet persistent or tardily deciduous .... **S. ×drummondii**
   2. Racemes fragile:
      4. Rhizome absent .... **S. arundinaceum**
      4. Rhizomes present:
         5. Sessile spikelets obtuse; glumes coriaceous; panicle 15–35 cm long, the branches 5–8 cm long; leaf-blades 8–15 mm wide .... **S. halepense**
         5. Sessile spikelets acute; glumes subcoriaceous with papery tip; panicle 20–60 cm long, the branches 15–20 cm long; leaf-blades 10–50 cm wide .... **S. propinquum**

S. arundinaceum (Desv.) Stapf in Prain, *Fl. Trop. Afr.* 9: 114 (1917). *Raphis arundinacea* Desv., *Opusc. Sci. Phys. Nat.*: 69 (1831). *Andropogon verticilliflorus* Steud., *Syn. Pl. Glumac.* 1: 393 (1854). *Sorghum verticilliflorum* (Steud.) Stapf in Prain, *Fl. Trop. Afr.* 9: 116 (1917); *Fl. Niue* 251; *Fl. Fiji* 376; *Fl. Sol.* 185.
COO, CRL, FIJ, HAW, NUE, NWC, PIT, SAM, SCI, SOL, TON; Old World tropics.

S. bicolor (L.) Moench, *Methodus*: 207 (1794); *Fl. Guam* 239; *Fl. Fiji* 376; *Man. Haw.* 1594; *Fl. Soc.* 345. *Holcus bicolor* L., *Mant. Pl.* 2: 301 (1771). *H. cafer* Ard., *Saggi Sci. Lett. Accad. Padova* 1: 119 (1786). *Sorghum vulgare* Pers., *Syn. Pl.*: 101 (1805); *Fl. N. Cal.* 27; *Fl. Niue* 252; *Pl. Samoa* 82; *Fl. Fiji* 376. *Andropogon subglabrescens* Steud., *Syn. Pl. Glumac.* 1: 393 (1854). *A. sorghum* var. *obovatus* Hack. in A. DC. & C. DC., *Monogr. Phan.* 6: 514 (1889). *A. sorghum* var. *subglobosus* Hack., l.c.: 515. *Sorghum miliiforme* var. *rotundulum* Snowden, *Bull. Misc. Inform., Kew* 1935: 237 (1935). *S. bicolor* var. *cafer* (Ard.) Fosberg & Sachet, *Micronesica* 18: 93 (1984); *List Micro.* 59. *S. bicolor* var. *obovatum* (Hack.) Fosberg & Sachet, l.c.: 94; *List Micro.* 59. *S. bicolor* var. *rotundulum* (Snowden) Fosberg & Sachet, l.c.: 94; *List Micro.* 59. *S. bicolor* var. *subglabrescens* (Steud.) Fosberg & Sachet, l.c.: 95; *List Micro.* 94.
FIJ, HAW, MRN, MRS, NUE, NWC, SCI, WAK; widely cultivated as a cereal crop.

S. ×drummondii (Nees ex Steud.) Millsp. & Chase, *Publ. Field Mus. Nat. Hist., Bot. Ser.* 3: 21 (1903). *Andropogon drummondii* Nees ex Steud., *Syn. Pl. Glumac.* 1: 393 (1854). *A. sorghum* var. *transiens* Hack. in A. DC. & C. DC., *Monogr. Phan.* 6: 508 (1889). *A. sorghum* subsp. *sudanensis* Piper, *Proc. Biol. Soc. Wash.* 28: 33 (1916). *Sorghum sudanense* (Piper) Stapf in Prain, *Fl. Trop. Afr.* 9:113 (1917); *Fl. N. Cal.* 27; *Fl. Pit.* 24; *List Micro.* 60. *S. vulgare* var. *sudanense* (Piper) Hitchc., *J. Wash. Acad. Sci.* 17: 147 (1927); *Pl. Samoa* 82. *S. bicolor* var. *transiens* (Hack.) Fosberg & Sachet, *Micronesica* 18: 96 (1984); *List Micro.* 60.
CRL, HAW, MRN, NWC, PIT, SAM, SCI; tropics & subtropics. This is a complex of hybrids between *Sorghum bicolor* and *S. arundinaceum*; a segregate from it is widely grown as a fodder crop under the name *S. sudanense*.

S. halepense (L.) Pers., *Syn. Pl.* 1: 101 (1805); *Pl. Tonga* 66; *Fl. Guam* 239; *Fl. Niue* 251; *Fl. Fiji* 375; *List Micro.* 60; *Fl. Sol.* 185; *Man. Haw.* 1594; *Easter* 82; *Fl. Soc.* 345. *Holcus halepensis* L., *Sp. Pl.* 2: 1047 (1753); *Fl. Raro.* 19. *Andropogon sorghum* subvar. *muticus* Hack. in A. DC. & C. DC., *Monogr. Phan.* 6: 502 (1889). *Sorghum halepense* f. *muticum* (Hack.) C. E. Hubb., *Hooker's Icon. Pl.* 34: t.3364 (1938); *Fl. Fiji* 375.
COO, CRL, EAS, FIJ, HAW, MRN, MRQ, NUE, NWC, PIT, SCI, SOL, TON; warm temperate & tropical regions.

S. nitidum (Vahl) Pers., *Syn. Pl.* 1: 101 (1805). *Holcus nitidus* Vahl., *Symb. Bot.* 2: 102 (1791).
CRL, MRN; tropical Asia.

S. propinquum (Kunth) Hitchc., *Lingnan Sci. J.* 7: 249 (1931). *Andropogon propinquus* Kunth, Enum. Pl. 1: 502 (1833). *Sorghum halepense* var. *propinquum* (Kunth) Ohwi, *Bot. Mag. (Tokyo)* 55: 550 (1941); *List Micro.* 60.
COO, CRL, MRN; Southeast Asia.

97. Chrysopogon

1. Racemes composed of several–many spikelet pairs .... **C. zizanioides**
1. Racemes reduced to a triad of 1 sessile and 2 pedicelled spikelets .... **C. aciculatus**

C. aciculatus (Retz.) Trin., *Fund Agrost.*: 188 (1820); *Fl. N. Cal.* 27; *Fl. Guam* 29; *Fl. Niue* 238; *Pl. Samoa* 87; *Fl. Fiji* 378; *List Micro.* 34; *Fl. Sol.* 183; *Man. Haw.* 1516; *Fl. Soc.* 336. *Andropogon aciculatus* Retz., *Observ. Bot.* 5: 22 (1789). *Rhaphis aciculata* (Retz.) Honda, *Bot. Mag. (Tokyo)* 11: 103 (1926); *Fl. Poly.* 54; *Fl. Raro.* 21; *Pl. Tonga* 66.
COO, CRL, FIJ, HAW, MRN, MRQ, MRS, NRU, NUE, NWC, SAM, SCI, SOL, TON, TUA, TUB, VAN, WAL; tropical Asia.

C. zizanioides (L.) Roberty, *Bull. Inst. Franç. Afrique Noire, Sér. A*, 22: 106 (1960). *Phalaris zizanioides* L., *Mant. Pl.* 2: 183 (1771). *Andropogon zizanioides* (L.) Urb., *Symb. Antill.* 4: 79 (1903); *Fl. Raro.* 17. *Vetiveria zizanioides* (L.) Nash in Small, *Fl. S.E. U.S.*: 67 (1903); *Fl. N. Cal.* 27; *Pl. Tonga* 53; *Fl. Niue* 253; *Fl. Fiji* 377; *List Micro.* 63; *Fl. Soc.* 348.
COO, CRL, FIJ, NUE, NWC, SAM, SCI, TON, TUB; tropics.

98. Dichanthium

1. Lower glume of sessile spikelet with a subapical arch of long hairs .... **D. sericeum**
1. Lower glume of sessile spikelet without a subapical arch:
   2. Peduncle pubescent above .... **D. aristatum**
   2. Peduncle glabrous:
      3. Lower glume of sessile spikelet narrowly oblong, with long bulbous-based hairs along margins above the middle .... **D. annulatum**
      3. Lower glume of sessile spikelet elliptic to obovate, glabrous to shortly ciliate on margins:
         4. Culms slender, erect, tufted; glumes glabrous or almost so .... **D. tenue**
         4. Culms robust, trailing or decumbent, rooting at lower nodes; glumes pilose below .... **D. caricosum**

D. annulatum (Forssk.) Stapf in Prain, *Fl. Trop. Afr.* 9: 178 (1917); *Fl. N. Cal.* 26; *Fl. Niue* 239; *Fl. Fiji* 384; *Man. Haw.* 1528. *Andropogon annulatus* Forssk., *Fl. Aegypt.-Arab.*: 173 (1775).
EAS, FIJ, HAW, MRQ, NRU, NUE, NWC, SOL, TON; Old World tropics, Central America.

D. aristatum (Poir.) C. E. Hubb., *Bull. Misc. Inform., Kew* 1939: 654 (1940); *Fl. Fiji* 384; *Man. Haw.* 1528. *Andropogon aristatus* Poir., *Encycl. Suppl.* 1: 585 (1810).
CRL, FIJ, HAW, MRN, MRQ, NWC, SOL; Old World tropics.

D. caricosum (L.) A. Camus, *Bull. Mus. Natl. Hist. Nat.* 27: 549 (1921); *Fl. N. Cal.* 26; *Fl. Guam* 230; *Fl. Niue* 239; *Fl. Fiji* 383; *List Micro.* 36; *Fl. Sol.* 184. *Andropogon caricosus* L., *Sp. Pl.*, ed. 2, 2: 1480 (1763).
FIJ, MRN, MRS, NUE, NWC, SOL; tropical Asia.

D. sericeum (R. Br.) A. Camus, *Bull. Mus. Natl. Hist. Nat.* 27: 549 (1921); *Fl. N. Cal.* 26; *Fl. Sol.* 184; *Man. Haw.* 1529. *Andropogon sericeus* R. Br., *Prodr.* 1: 201 (1810).
HAW, MRQ, NWC, SOL; New Guinea, Philippines, Australia.

D. tenue (R. Br.) A. Camus, *Bull. Mus. Natl. Hist. Nat.* 27: 549 (1921); *Man. Haw.* 1528. *Andropogon tenuis* R. Br., *Prodr.* 1: 201 (1810).
HAW, TON; Australia.

99. Capillipedium

1. Culms decumbent, branched; nodes glabrous; sessile spikelets 1.8–3 mm long, the pedicelled about as long .... **C. assimile**
1. Culms erect, not or sparsely branched; nodes pubescent; sessile spikelets 3–4 mm long, the pedicelled usually shorter .... **C. spicigerum**

C. assimile (Steud.) A. Camus in Lecomte, *Fl. Indo-Chine* 7: 314 (1922). *Andropogon assimilis* Steud., *Syn. Pl. Glumac.* 1: 397 (1854).
NRU; tropical Asia.

C. spicigerum S. T. Blake, *Pap. Dept. Biol. Univ. Queensland* 2(3): 43 (1944); *Fl. N. Cal.* 26; *Fl. Sol.* 183. *C. parviflorum sensu Fl. N. Cal.* 27.
CRL, NWC, SOL; Southeast Asia, Australia.

100. Bothriochloa

1. Pedicelled spikelets linear, much narrower than the sessile:
   2. Culm nodes pubescent; sessile spikelets 4.5–7.3 mm long . . . . **B. barbinodis**
   2. Culm nodes glabrous; sessile spikelets 2.8–3.5 mm long . . . . **B. laguroides**
1. Pedicelled spikelets elliptic, about as wide as the sessile:
   3. Racemes borne on a central axis longer than themselves . . . . **B. bladhii**
   3. Racemes subdigitate or with a central axis shorter than the lowest raceme:
      4. Lower glume of sessile spikelets pitted . . . . **B. pertusa**
      4. Lower glume of sessile spikelets without pits . . . . **B. ischaemum**

B. barbinodis (Lag.) Herter, *Revista Sudamer. Bot.* 6: 135 (1940); *Man. Haw.* 1502. *Andropogon barbinodis* Lag., *Gen. Sp. Pl.*: 3 (1816).
HAW, MRQ; southern USA to Argentina.

B. bladhii (Retz.) S. T. Blake, *Proc. Roy. Soc. Queensland* 80: 62 (1969); *Fl. Fiji* 385. *Andropogon bladhii* Retz., *Observ. Bot.* 2: 27 (1781). *A. intermedius* R. Br., *Prodr.* 1: 202 (1810); *Pl. Tonga* 65. *Amphilophis intermedia* (R. Br.) Stapf in Prain, *Fl. Trop. Afr.* 9: 174 (1917); *Fl. N. Cal.* 26. *Dichanthium bladhii* (Retz.) Clayton, *Kew Bull.* 32: 3 (1977); *List Micro.* 36.
CRL, FIJ, HAW, MRN, MRQ, MRS, NRU, NWC, SAM, SCI, TON; Old World tropics.

B. ischaemum (L.) Keng, *Contr. Biol. Lab. Chin. Assoc. Advancem. Sci., Sect. Bot.* 10: 201 (1936); *Easter* 73. *Andropogon ischaemum* L., *Sp. Pl.* 2: 1047 (1753).
EAS; warm temperate Old World.

B. laguroides (DC.) Herter, *Revista Sudamer. Bot.* 6: 135 (1940). *Andropogon laguroides* DC., *Cat. Pl. Horti Monsp.*: 78 (1813).
HAW; Central & South America.

B. pertusa (L.) A.Camus, *Ann. Soc. Linn. Lyon, n.s.* 76: 164 (1931); *Man. Haw.* 1503. *Holcus pertusus* L., *Mant. Pl.* 2: 301 (1771). *Amphilophis pertusa* (L.) Nash ex Stapf, *Agric. News (Barbados)* 15: 179 (1916); *Fl. N. Cal.* 26.
FIJ, HAW, MRN, MRQ, NWC, SAM, TON, WAL; tropical Asia.

101. Ischaemum

1. Sessile spikelet lower glume rugose on back or flanks:
   2. Lower glume coarsely rugose across the back; annual . . . . **I. rugosum**
   2. Lower glume knobbly on the flanks, rarely extending to weak transverse ridges; perennial . . . . **I. barbatum**
1. Sessile spikelet lower glume smooth:

3. Pedicelled spikelet awnless, the sessile with or without a weak awn 1–12 mm long; lower glume of sessile spikelet winged above:
 4. Inflorescence enclosed at the base by uppermost sheath; rhizomatous . . **I. muticum**
 4. Inflorescence exserted from uppermost sheath; loosely tufted . . . . **I. aristatum**
3. Pedicelled and sessile spikelet both geniculately awned, the awn 5–35 mm long:
 5. Racemes (2 –)3–8 . . . . **I. polystachyum**
 5. Racemes strictly paired:
  6. Lower glume of sessile spikelet distinctly winged above; upper glume winged, with or without a brief awn (1–2 mm) . . . . **I. setaceum**
  6. Lower glume of sessile spikelet not, or obscurely, winged:
   7. Back of rhachis with a basal circular pore between internode and pedicel . . . . **I. timorense**
   7. Back of rhachis without a distinct circular pore:
    8. Awn of fertile lemma 30–35 mm long; sessile spikelet 7–9 mm long; glumes of both spikelets awnless . . . . **I. longisetum**
    8. Awn of fertile lemma 15–25 mm long; lower glume of sessile spikelet usually 2-awned (0.5–2 mm), the upper 1-awned (1–5 mm); pedicelled spikelet glumes awned (0.5–4mm):
     9. Sessile spikelet 4–6 mm long, its lower glume glabrous . . **I. murinum**
     9. Sessile spikelet 6–7 mm long, its lower glume hirsute . . . . . . **I. byrone**

I. aristatum L., *Sp. Pl.* 2: 1049 (1753); *Fl. Sol.* 184. *I. indicum sensu Fl. Niue* 243; *Fl. Fiji* 273; List Micro. 46.
CRL, FIJ, HAW, NUE, SAM, SOL; Southeast Asia.

I. barbatum Retz., *Observ. Bot.* 6: 35 (1791); *List Micro.* 46. *I. vitiense* Summerh., *Bull. Misc. Inform., Kew* 1930: 253, 264 (1930); *Fl. Fiji* 274.
FIJ, MRN; tropical Asia.

I. byrone (Trin.) Hitchc., *Mem. Bernice Puauhi Bishop Mus.* 8: 213 (1922); *Man. Haw.* 1556. *Spodiopogon byronis* Trin., *Mém. Acad. Imp. Sci. St.-Pétersbourg, Sér. 6, Sci. Math.* 2: 301 (1833). *Andropogon byronis* (Trin.) Steud., *Syn. Pl. Glumac.* 1: 398 (1854). *Ischaemum lutescens* Hack. in A. DC. & C. DC., *Monogr. Phan.* 6: 221 (1889).
FIJ, HAW, NUE, NWC, SAM, SCI, TON.

I. longisetum Merr., *Philipp. J. Sci., Bot.* 9: 52 (1914); *Fl. Guam* 234; *List Micro.* 46. *I. longisetum* var. *raulersoniae* Fosberg & Sachet, *Micronesica* 18: 62 (1984); *List Micro.* 46.
MRN.

I. murinum G. Forst., *Nova Acta Regiae Soc. Sci. Upsal.* 3: 185 (1780); *Fl. N. Cal.* 26; *Pl. Tonga* 67. *Andropogon murinus* (G. Forst.) Steud., *Syn. Pl. Glumac.* 1: 377 (1854). *Ischaemum foliosum* Hack. in A. DC & C. DC., *Monogr. Phan.* 6: 222 (1889); *Fl. Niue* 243. *I. foliosum* var. *leiophyllum* Ridl., *J. Straits Branch Roy. Asiat. Soc.* 45: 243 (1906). *I. stokesii* F. Br., *Bernice P. Bishop Mus. Bull.* 84: 53 (1931); *Fl. Poly.* 53. *I. littorale* Reeder, *J. Arnold Arbor.* 29: 345 (1948); *Fl. Sol.* 184.
COO, NUE, NWC, SAM, SCI, SOL, TON, TUB, VAN; New Guinea.

I. muticum L., *Sp. Pl.* 2: 1049 (1753); *Fl. N. Cal.* 26; *List Micro.* 46; *Fl. Sol.* 184. *I. muticum* var. *aristuliferum* Fosberg & Sachet, *Micronesica* 18: 64 (1984); *List Micro.* 46.
CRL, NWC, SOL, VAN; Southeast Asia.

I. polystachyum J. Presl in C. Presl, *Reliq. Haenk.* 1: 828 (1830); *List Micro.* 46. *I. intermedium* Brongn. in Duperrey, *Voy. Monde Phan.*: 73 (1831); *Fl. N. Cal.* 26. *Spodiopogon chordatus* Trin., *Mém. Acad. Imp. Sci. St.-Pétersbourg, Sér. 6, Sci. Math.* 2: 302 (1833). *Andropogon chordatus* (Trin.) Steud., *Syn. Pl. Glumac.* 1: 398 (1854). *A. mariannae* Steud., l.c.: 382 (1854). *A. paniceus* Steud., l.c.: 375. *Ischaemum digitatum* var. *polystachyum* (J. Presl) Hack. in A. DC. & C. DC., *Monogr. Phan.* 6: 233 (1889); *Fl. Guam* 234. *I. chordatum* (Trin.) Hack. ex Warb., *Bot. Jahrb. Syst.* 13: 260 (1891). *I. polystachyum* var. *chordatum* (Trin.) Fosberg & Sachet, *Micronesica* 18: 66 (1984); *List Micro.* 46. *I. polystachyum* var. *hillii* Fosberg & Sachet, l.c.: 67; *List Micro.* 47. *I. polystachyum* var. *intermedium* (Brongn.) Fosberg & Sachet, l.c.: 67; *List Micro.* 47.
CRL, MRN, NWC, SOL, VAN; Old World tropics.

I. rugosum Salisb., *Icon. Stirp. Rar.* 1: t.1 (1791); *Fl. Guam* 234; *Fl. Fiji* 372. *I. segetum* Trin., *Mém. Acad. Imp. Sci. St.-Pétersbourg, Sér. 6, Sci. Math.* 2: 294 (1833). *I. rugosum* var. *segetum* (Trin.) Hack. in A. DC & C. DC., *Monogr. Phan.* 6: 208 (1889); *List Micro.* 47.
CRL, FIJ, MRN; tropics.

I. setaceum Honda, *Bot. Mag. (Tokyo)* 38: 54 (1924); *List Micro.* 47.
CRL; Taiwan.

I. timorense Kunth, *Révis. Gramin.* 1: 369 (1830); *Fl. Fiji* 273; *List Micro.* 47.
CRL, FIJ, HAW; tropical Asia.

102. Apluda

A. mutica L., *Sp. Pl.* 1: 82 (1753); *List Micro.* 31; *Fl. Sol.* 183. *A. varia* var. *mutica* (L.) Hack. in A. DC. & C. DC., *Monogr. Phan.* 6: 197 (1889); *Fl. N. Cal.* 26.
CRL, NWC, SOL, VAN; tropical Asia.

103. Dimeria

1. Perennial, 60–90 cm high; leaf-blades 10–15 cm long; spikelets 3.3–4.8 mm long .... **D. chloridiformis**
1. Annual, 20–50 cm high; leaf-blades 2–5 cm long; spikelets 0.8–3 mm long:
   2. Racemes paired, glabrous on rhachis margins .... **D. ornithopoda**
   2. Racemes 3–4, digitate, ciliate on rhachis margins .... **D. paniculata**

D. chloridiformis (Gaudich.) K. Schum. & Lauterb., *Fl. Schutzgeb. Südsee*: 165 (1901); *Fl. Guam* 231; *List Micro.* 40. *Andropogon chloridiformis* Gaudich., *Voy Uranie*: 412 (1830). *Dimeria ciliata* var. *heteromorpha* Reeder, *J. Arnold Arbor.* 29: 325 (1948); *List Micro.* 40.
CRL, MRN; Malaya, Philippines, New Guinea, Australia.

D. ornithopoda Trin., *Fund. Agrost.*: 167 (1820); *Fl. Guam* 231. *D. tenera* Trin., *Mém. Acad. Imp. Sci. St.-Pétersbourg, Sér. 6, Sci. Math.* 2: 335 (1833). *D. ornithopoda* var. *tenera* (Trin.) Hack. in A. DC. & C. DC., *Monogr. Phan.* 6: 80 (1889); *Fl. Guam* 231; *List Micro.* 40.
CRL, MRN; tropical Asia, Australia.

D. paniculata Masam., *Trans. Nat. Hist. Soc. Taiwan* 28: 149 (1938).
CRL.

104. Andropogon

1. Pedicelled spikelet male, as large as the sessile spikelet .... **A. gerardii**
1. Pedicelled spikelet vestigial or represented by the pedicel alone:
   2. Raceme-pairs profuse and crowded into an oblong or corymbose mass .... **A. glomeratus**
   2. Raceme-pairs more or less separate in a loose linear false panicle ..... **A. virginicus**

A. gerardii Vitman, *Summa Pl.* 6: 16 (1792).
HAW; North America.

A. glomeratus (Walter) Britton, Sterns & Poggenb., *Prelim. Cat.*: 67 (1888). *Cinna glomerata* Walter, *Fl. Carol.*: 59 (1788).
HAW; southern USA, Mexico, Caribbean.

A. virginicus L., *Sp. Pl.* 2: 1046 (1753); *Man. Haw.* 1497.
HAW; southern USA, Mexico, Caribbean.

105. Cymbopogon

1. Sessile spikelet awnless:
   2. Pedicelled spikelet reduced to a single glume, this pungent .... **C. refractus**
   2. Pedicelled spikelet male, acute .... **C. citratus**
1. Sessile spikelet awned:
   3. Lowest pedicel of sessile raceme swollen .... **C. coloratus**
   3. Lowest pedicel of sessile raceme not swollen .... **C. nardus**

C. citratus (DC.) Stapf, *Bull. Misc. Inform., Kew* 1906: 357 (1906); *Fl. Raro.* 18; *Fl. N. Cal.* 27; *Fl. Guam* 230; *Pl. Samoa* 83; *List Micro.* 35; *Fl. Soc.* 332. *Andropogon citratus* DC., *Cat. Pl. Horti Monsp.*: 78 (1813).
COO, CRL, FIJ, HAW, MRN, MRS, NWC, SAM, SCI, TON; widely cultivated as lemon grass. The species seldom flowers.

C. coloratus (*Hook. f.*) *Stapf*, *Bull. Misc. Inform., Kew* 1906: 321 (1906); *Fl. Fiji* 379; *Fl. Sol.* 184. *Andropogon nardus* var. *coloratus* Hook. f., *Fl. Brit. Ind.* 7: 206 (1896).
FIJ, SOL; tropical Asia.

C. nardus (L.) Rendle in Hiern, *Cat. Afr. Pl.* 2: 155 (1899); *Fl. Guam* 230; *Fl. Niue* 238; *List Micro.* 35. *Andropogon nardus* L., *Sp. Pl.* 2: 1046 (1753).
CRL, FIJ, MRN, NUE; Old World tropics.

C. refractus (R. Br.) A. Camus, *Rev. Bot. Appl. Agric. Colon.* 1: 270 (1921); *Fl. N. Cal.* 27; *Pl. Tonga* 66; *Fl. Fiji* 379; *Man. Haw.* 1519; *Fl. Soc.* 332. *Andropogon refractus* R. Br., *Prodr.* 1: 202 (1810); *Fl. Poly.* 52. *A. tahitensis* Hook. & Arn., *Bot. Beechey Voy.*: 72 (1841).
COO, FIJ, HAW, MRN, NWC, SCI, TON, TUA, TUB, VAN; Australia.

106. Schizachyrium

1. Spikelets divergent from a more or less zigzag rhachis with curved pedicels; tufted perennials:
    2. Inflorescence profusely branched to form a dense head of crowded racemes; rhachis and pedicels plumose; sessile spikelets 3.5–5 mm long .... **S. condensatum**
    2. Inflorescence loose, the racemes separate; rhachis and pedicels ciliate; sessile spikelets 6–8 mm long .... **S. scoparium**
1. Spikelets and pedicels appressed to a straight rhachis; annuals:
    3. Lower glume of sessile spikelet winged; rhachis internodes with a dense beard of long hairs .... **S. fragile**
    3. Lower glume of sessile spikelet without wings; rhachis internodes glabrous to ciliate:
        4. Spatheole convolute, barely reaching base of raceme; sessile spikelets 2–4 mm long; awn 11–30 mm long; leaf-blades broadly obtuse .... **S. brevifolium**
        4. Spatheole lanceolate, more or less enclosing raceme; sessile spikelet 4–8 mm long; awn 7–11 mm long; leaf-blades acute to narrowly obtuse ..... **S. pseudeulalia**

S. brevifolium (Sw.) Büse in Miq., *Pl. Jungh.*: 359 (1854); *Fl. Guam* 232. *Andropogon brevifolius* Sw., *Prodr.*: 26 (1788). *Schizachyrium paradoxum* Büse in Miq., *Pl. Jungh.*: 359 (1854). *Andropogon brevifolius* var. *paradoxus* (Büse) Ohwi, *Bot. Mag. (Tokyo)* 55: 550 (1941); *List Micro.* 30.
CRL, MRN; throughout the tropics.

S. condensatum (Kunth) Nees, *Agrost. Bras.*: 333 (1829); *Man. Haw.* 1590. *Andropogon condensatus* Kunth in Humb. & Bonpl., *Nov. Gen. Sp.* 1: 188 (1815).
HAW; tropical America.

S. fragile (R. Br.) A. Camus, *Ann. Soc. Linn. Lyon, n.s.* 70: 87 (1924); *Fl. Guam* 232. *Andropogon fragilis* R. Br., *Prodr.* 1: 202 (1810); *List Micro.* 30. *A. obliquiberbis* Hack., *Flora* 68: 117 (1885). *Schizachyrium obliquiberbe* (Hack.) A. Camus, *Ann. Soc. Linn. Lyon, n.s.* 70: 89 (1924); *Fl. N. Cal.* 26.
CRL, MRN, NWC; Southeast Asia, Australia.

S. pseudeulalia (Hosok.) S. T. Blake, *Proc. Roy. Soc. Queensland* 80: 77 (1969). *Microstegium pseudeulalia* Hosok., *Trans. Nat. Hist. Soc. Taiwan* 28: 151 (1938).
CRL; Malesia, Australia.

S. scoparium (Michx.) Nash in Small, *Fl. S.E. U.S.*: 59 (1903). *Andropogon scoparius* Michx., *Fl. Bor.-Amer.* 1: 57 (1803).
HAW; North America.

107. Arthraxon

A. hispidus (Thunb.) Makino, *Bot. Mag. (Tokyo)* 27: 214 (1912). *Phalaris hispida* Thunb. in Murray, *Syst. Veg.*, ed. 14: 104 (1784).
HAW; Old World tropics.

108. Hyparrhenia

1. Spikelets with red hairs .... **H. rufa**
1. Spikelets with white hairs or glabrous:
   2. Base of individual racemes in the pair filiform, unequal, glabrous or softly hairy .... **H. hirta**
   2. Base of individual racemes in the pair flattened, subequal, stiffly setose ... **H. dregeana**

H. dregeana (Nees) Stapf ex Stent, *Bothalia* 1: 249 (1923). *Andropogon dregeanus* Nees, *Fl. Afr. Austral. Ill.* 1: 112 (1841).
HAW; eastern & southern Africa.

H. hirta (L.) Stapf in Prain, *Fl. Trop. Afr.* 9: 315 (1918); *Man. Haw.* 1554. *Andropogon hirtus* L., *Sp. Pl.* 2: 1046 (1753).
HAW; Old World tropics & subtropics.

H. rufa (Nees) Stapf in Prain, *Fl. Trop. Afr.* 9: 304 (1918); *Fl. Fiji* 380; *List Micro.* 44; *Man. Haw.* 1554; *Fl. Soc.* 338. *Trachypogon rufus* Nees, *Agrost. Bras.*: 345 (1829).
FIJ, HAW, MRN, SCI; Africa and widely introduced.

109. Heteropogon

H. contortus (L.) P. Beauv. ex Roem. & Schult., *Syst. Veg.* 2: 836 (1817); *Fl. N. Cal.* 27; *Pl. Tonga* 67; *Fl. Guam* 233; *Fl. Fiji* 381; *List Micro.* 44; *Man. Haw.* 1550; *Fl. Soc.* 338. *Andropogon contortus* L., *Sp. Pl.*, ed. 2, 2: 1045 (1763); *Fl. Poly.* 52.
FIJ, HAW, MRN, MRQ, NWC, SAM, SCI, TON; throughout the tropics.

110. Themeda

1. Sessile spikelets prominently awned; pairs of involucral spikelets inserted at the same level:
   2. Involucral spikelets 4.5–6 mm long; sessile spikelet 5–5.5 mm long; awn 30–45 mm long .... **T. quadrivalvis**
   2. Involucral spikelets 6–14 mm long; sessile spikelet 6–11 mm long:
      3. Racemes in flattened fan-like fascicles; involucral spikelets reduced to one papery glume; pedicelled spikelets linear, without lemmas; awns 50–70 mm long; annual .... **T. arguens**

3. Racemes in irregular clumps; involucral spikelets well developed, usually male; pedicelled spikelets lanceolate, male; awns 25–70 mm long; perennial .... **T. triandra**

1. Sessile spikelets awnless or with an inconspicuous awn up to 10 mm long; pairs of involucral spikelets separated by a brief internode:
   4. Involucral spikelets glabrous .... **T. villosa**
   4. Involucral spikelets densely hirsute with fulvous tubercle-based hairs:
      5. Racemes containing 1 sessile spikelet, 1 cm long, falling entire; spatheoles 1–1.5 cm long; involucral spikelets 6.5–8 mm long .... **T. gigantea**
      5. Racemes containing 2–3 sessile spikelets, 1.5–2 cm long, shedding spikelets at maturity; spatheoles 2–3.5 cm long; involucral spikelets 10–12 mm long .... **T. intermedia**

T. arguens (L.) Hack. in A. DC. & C. DC., *Monogr. Phan.* 6: 657 (1889); *Fl. Fiji* 382. *Stipa arguens* L., *Sp. Pl.*, ed. 2, 1: 117 (1762).
FIJ; Southeast Asia.

T. gigantea (Cav.) Hack. in Duthie, *Fodder Grass. N. Ind.*: 89 (1888); *Fl. N. Cal.* 26; *Fl. Sol.* 186. *Anthistiria gigantea* Cav., *Icon.* 5: 36 (1799).
NWC, SOL, VAN; Vietnam, Borneo, New Guinea, Philippines.

T. intermedia (Hack.) Bor, *Indian Forest Rec., Bot.* 1: 96 (1938). *T. gigantea* var. *intermedia* Hack. in A. DC. & C. DC., *Monogr. Phan.* 6: 675 (1889).
SOL, VAN; tropical Asia.

T. quadrivalvis (L.) Kuntze, *Revis. Gen. Pl.* 2: 794 (1891); *Fl. Fiji* 382. *Andropogon quadrivalvis* L., *Syst. Veg.*, ed. 13,: 758 (1774). *Anthistiria ciliata* L. f., *Suppl. Pl.*: 113 (1781), nom. superfl. *Themeda ciliata* Hack. in A. DC. & C. DC., *Monogr. Phan.* 6: 661 (1889); *Fl. N. Cal.* 26.
FIJ, NWC, SOL; tropical Asia.

T. triandra Forssk., *Fl. Aegypt.-Arab.*: 178 (1775); *Fl. N. Cal.* 26. *Anthistiria australis* R. Br., *Prodr.* 1: 200 (1810). *Themeda australis* (R. Br.) Stapf in Prain, *Fl. Trop. Afr.* 9: 420 (1919); *Fl. Sol.* 186.
NWC, SOL; Old World tropics.

T. villosa (Poir.) A. Camus in Lecomte, *Fl. Indo-Chine* 7: 364 (1922); *Man. Haw.* 1599. *Anthistiria villosa* Poir., *Encycl. Suppl.* 1: 396 (1812).
HAW; tropical Asia.

111. Elionurus

E. citreus (R. Br.) Benth., *Fl. Austral.* 6: 510 (1878); *Fl. Sol.* 184. *Andropogon citreus* R. Br., *Prodr.* 1: 203 (1810).
SOL; New Guinea, Australia.

112. Eremochloa

E. ophiuroides (Munro) Hack. in A. DC. & C. DC., *Monogr. Phan.* 6: 261 (1889); *Fl. Guam* 233; *List Micro.* 43. *Ischaemum ophiuroides* Munro, *Proc. Amer. Acad. Arts* 4: 363 (1860).
HAW, MRN; Southeast Asia and often introduced as a lawn grass.

113. Rottboellia

1. Leaf-blades linear, 5–20 mm wide, tapering to the base .... **R. cochinchinensis**
1. Leaf-blades lanceolate, 25–35 mm wide, rounded to cordate at the base .. **R. coelorachis**

R. cochinchinensis (Lour.) Clayton, *Kew Bull.* 35: 817 (1981); *Fl. Sol.* 185. *R. exaltata* L.f., Suppl. Pl.: 114 (1781), non L. f. (1779). *Stegosia cochinchinensis* Lour., *Fl. Cochinch.*: 51 (1790).
SOL; Old World tropics.

R. coelorachis G. Forst., *Fl. Ins. Austr.*: 9 (1786); *Fl. N. Cal.* 26.
NWC; Norfolk Is.

114. Hackelochloa

H. granularis (L.) Kuntze, *Revis. Gen. Pl.* 2: 776 (1891); *List Micro.* 43. *Cenchrus granularis* L., *Mant. Pl.* 2, App.: 575 (1771).
CRL, HAW, SOL; tropics.

115. Mnesithea

M. laevis (Retz.) Kunth, *Révis. Gramin.* 1: 154 (1829). *Rottboellia laevis* Retz., *Observ. Bot.* 3: 11 (1783). *Phleum cochinchinense* Lour., *Fl. Cochinch.*: 48 (1790). *Heteropholis cochinchinensis* (Lour.) Clayton, *Kew Bull.* 35: 816 (1981); *List Micro.* 44.
CRL, MRN; tropical Asia.

116. Hemarthria

1. Lower glume of sessile spikelet 4–6 mm long .... **H. altissima**
1. Lower glume of sessile spikelet 3–4 mm long .... **H. compressa**

H. altissima (Poir.) Stapf & C. E. Hubb., *Bull. Misc. Inform., Kew* 1934: 109 (1934). *Rottboellia altissima* Poir., *Voy. Barbarie* 2: 105 (1789).
HAW; Old World tropics.

H. compressa (L. f.) R. Br., *Prodr.*: 207 (1810); *Fl. Sol.* 184. *Rottboellia compressa* L.f., *Suppl. Pl.*: 114 (1781).
SOL; tropical Asia.

117. Tripsacum

1. Male spikelets 6–10 mm long, acute, one of the pair sessile, the other on a pedicel 1–3 mm long .... **T. andersonii**
1. Male spikelets 3–5 mm long, obtuse, both sessile or one on a pedicel up to 0.4 mm long .... **T. latifolium**

T. andersonii J. R. Gray, *Phytologia* 33: 204 (1976). *T. laxum sensu Fl. Niue* 253: *Fl. Fiji* 389; *List Micro*. 62; *Fl. Sol*. 186; *Fl. Soc*. 347.
FIJ, MRN, NUE, SCI, SOL; Central America and widely introduced as a fodder crop. Commonly misnamed as *Tripsacum laxum*.

T. latifolium Hitchc., *Bot. Gaz*. 41: 294 (1906); *Fl. Guam* 240; *List Micro*. 62.
MRN; Central America.

118. Zea

Z. mays L., *Sp. Pl*. 2: 971 (1753); *Fl. N. Cal*. 24; *Fl. Guam* 241; *Fl. Niue* 254; *Pl. Samoa* 108; *Fl. Fiji* 387; *Fl. Pit*. 24; *List Micro*. 63.
CRL, FIJ, HAW, MRN, MRS, NUE, NWC, PIT, SAM, VAN, WAK; cultivated in the tropics and subtropics as maize or corn.

119. Chionachne

C. macrophylla (Benth.) Clayton, *Kew Bull*. 35: 814 (1981). *Polytoca macrophylla* Benth., *J. Linn. Soc. Bot*. 19: 52 (1881); *Fl. Sol*. 185
HAW, SOL; Malesia.

120. Coix

C. lacryma-jobi L., *Sp. Pl*. 2: 972 (1753); *Fl. Poly*. 51; *Fl. Raro*. 17; *Fl. N. Cal*. 24; *Pl. Tonga* 67; *Fl. Guam* 140; *Fl. Niue* 238; *Pl. Samoa* 108; *Fl. Fiji* 388; *Fl. Pit*. 23; *List Micro*. 34; *Fl. Sol*. 184; *Man. Haw*. 1617; *Easter* 74; *Fl. Soc*. 332.
COO, CRL, EAS, FIJ, HAW, MRN, MRQ, NUE, NWC, PIT, SAM, SCI, SOL, TON, TUB, VAN; tropical Asia and widely introduced for its ornamental fruit.

Index

Accepted names in **bold** type; synonyms in *italic*.

Agropyron repens 34
Agrostis 30
africana 47
avenacea 30, 31
canina 31
capillaris 31
diandra 47
exarata 31
fallax 31
fertilis 47
filiformis 31
indica 47
matrella 51
miliacea 25
procera 62
pyramidata 48
radiata 49
rapensis 32
rockii 31
sandwicensis 31
semiverticillata 33
stolonifera 31
ventricosa 32
verticillata 33
virginica 48
viridis 33
Aira 30
australis 29
caryophyllea 30
hawaiiensis 29
f. *depauperata* 29
f. *haleakalensis* 29
indica 58
macrantha 29
nubigena 29
pallida 29
var. *tenuissima* 29
Alloteropsis 59
semialata 59
Alopecurus 33
anthoxanthoides 33
monspeliensis 33
utriculatus 33
Ammophila 32
arenaria 32
Amphilophis intermedia 86
pertusa 86
Ancistrachne 57
numaeensis 3, 57
uncinulata 57
Andropogon 89
aciculatus 84
amaurus 82
annulatus 85
aristatus 85
assimilis 85
aureus 82
barbatus 48
barbinodis 86
bladhii 86
brevifolius 90
var. *paradoxus* 90
byronis 87
caricosus 85
chloridiformis 88
chordatus 88
citratus 89
citreus 92
condensatus 90
contortus 91
crinitus 82
dregeanus 91
drummondii 84
fragilis 90
gerardii 89
glomeratus 89
hirtus 91
insularis 72
intermedius 86
ischaemum 86
laguroides 86
mariannae 88
murinus 87
nardus 90
var. *coloratus* 89
obliquiberbis 90

paniceus 88
propinquus 84
quadrivalvis 92
refractus 90
scoparius 91
sericeus 85
sorghum subsp. *sudanensis* 84
subvar. *muticus* 84
var. *obovatus* 83
var. *subglobosus* 83
var. *transiens* 84
subglabrescens 83
tahitensis 90
tenuis 85
verticilliflorus 83
virginicus 89
zizanioides 84
ANDROPOGONEAE 4, 80
Anthephora 76
hermaphrodita 76
Anthistiria australis 92
ciliata 92
gigantea 92
villosa 92
Anthoxanthum 30
crinitum 32
odoratum 30
Apluda 88
mutica 88
varia var. *mutica* 88
Aristida 37
adscensionis 37
aspera 38
novae-caledoniae 37, 38
pilosa 37, 38
ramosa 37, 38
repens 37, 38
ARISTIDEAE 4, 37
Arrhenatherum 28
elatius 28
Arthraxon 91
hispidus 91
ARUNDINEAE 4, 35
ARUNDINELLEAE 4, 78
Arundo 36
arenaria 32
australis 37
donax 36
var. *versicolor* 36
karka 37
selloana 36
semiannularis 36
vallatoria 37
versicolor 36
Austrodanthonia biannularis 36
Avena 28
barbata 28
elatior 28
fatua 28
filiformis 31
flavescens 29
sativa 28
AVENEAE 4, 28
Axonopus 65
affinis 65
compressus 65
fissifolius 65
paschalis 65

Bambusa arundinacea 3
balcooa 3
beecheyana 3
blumeana 3
multiplex 3
oldhamii 3
solomonensis 3
tuldoides 3
vulgaris 3
BAMBUSEAE 4
Bothriochloa 86
barbinodis 86
bladhii 86
ischaemum 86
laguroides 86
pertusa 86
Brachiaria ambigua 61
brizantha 60
decumbens 60
distachya 60
eruciformis 60
humidicola 61
miliiformis 62
mollis 61
mutica 61
paspaloides 61
plantaginea 61
reptans 61
subquadripara 62
Briza 26
maxima 26, 27

minor 26, 27
BROMEAE 4, 33
Bromus 33
catharticus 33, 34
diandrus 33, 34
var. *rigidus* 34
hordeaceus 33, 34
madritensis 33, 34
mollis 34
rigidus 34
rubens 33, 34
secalinus 33, 34
sterilis 33, 34
tectorum 33, 34
unioloides 34
willdenowii 34

Calamagrostis 31
expansa 3, 31
hillebrandii 3, 31, 32
Capillipedium 85
assimile 85
parviflorum 86
spicigerum 85, 86
Cenchrus 75
agrimonioides 3, 75
var. *laysanensis* 75
anomoplexis 76
australis var. *latifolius* 76
brownii 75
caliculatus 3, 75, 76
var. *uniflorus* 75
ciliaris 75, 76
echinatus 75, 76
var. *glabratus* 76
var. *hillebrandianus* 76
var. *pennisetoides* 76
fusiformis 75
granularis 93
hillebrandianus 76
laniflorus 76
lappaceus 35
laysanensis 75
parviflorus 66
pedunculatus 75
setigerus 75, 76
setosus 74
taitensis 76
tribuloides 75, 76
Centotheca 35
lappacea 35
latifolia 35
malabarica 45
CENTOTHECEAE 4, 35
Chaetochloa verticillata 67
Chionachne 94
macrophylla 94
Chloris 48
barbata 48
cheesemanii 3, 49
cynodontoides 49
divaricata 48, 49
var. *cynodontoides* 49
dolichostachya 49
gayana 48, 49
incompleta 49
inflata 48
petraea 49
radiata 48, 49
truncata 48, 49
unispicea 49
virgata 48, 49
Chrysopogon 84
aciculatus 84
zizanioides 84
Cinna glomerata 89
Coix 94
lacryma-jobi 94
Cortaderia 36
jubata 36
selloana 36
Cymbopogon 89
citratus 89
coloratus 89
nardus 89, 90
refractus 89, 90
Cynodon 50
aethiopicus 50
dactylon 50
var. *maritimus* 50
var. *parviglumis* 50
maritimus 50
nlemfuensis 50
parviglumis 50
CYNODONTEAE 4, 48
Cynosurus aegyptius 46
aureus 26
coracanus 46
indicus 46
virgatus 45

Cyrtococcum 58
accrescens 58
oxyphyllum 58
patens 58
trigonum 58

Dactylis 28
glomerata 28
Dactyloctenium 46
aegyptium 46
var. *radicans* 46
Danthonia paschalis 36
pilosa 36
semiannularis 36
Deschampsia 29
australis 29
f. *haleakalensis* 29
subsp. *nubigena* 29
var. *gracilis* 29
var. *tenuissima* 29
hawaiiensis 29
klossii 29
nubigena 29
pallens 29
Deyeuxia expansa 31
hillebrandii 32
Dichanthelium conjugens 55
cynodon 55
forbesii 55
hillebrandianum 55
isachnoides 55
koolauense 56
Dichanthium 85
annulatum 85
aristatum 85
bladhii 86
caricosum 85
sericeum 85
tenue 85
Dichelachne 32
crinita 32
micrantha 32
Digitaria 69
abyssinica 69, 70
adscendens 71
bicornis 70, 71
caledonica 69, 71
ciliaris 70, 71
collina 72
decumbens 71
didactyla 70, 71
divaricatissima 69, 71
eriantha 70, 71
filiformis 70, 71
fuscescens 69, 71
gaudichaudii 69, 71
henryi 71
heterantha 69, 71
horizontalis 70, 72
insularis 69, 72
latronum 72
longiflora 69, 72
longissima 71
marianensis 72
mariannensis 69, 72
mezii 70, 72
microbachne 72
subsp. *calliblepharata* 72
montana 70, 72
nuda 70, 72
pacifica 73
pentzii 71
propinqua 72
pruriens 72
var. *microbachne* 72
radicosa 70, 72
robinsonii 71
sanguinalis 71
setigera 70, 72
var. *calliblepharata* 73
stenotaphrodes 69, 73
stricta 71
timorensis 72
violascens 69, 73
Dimeria 88
chloridiformis 88
ciliata var. *heteromorpha* 88
ornithopoda 88, 89
var. *tenera* 89
paniculata 88, 89
tenera 89
Diplax avenacea 24
Dissochondrus 66, 68
bifidus 68
biflorus 3, 68
Distichlis 38
spicata 38

Echinochloa 58
colona 58, 59

crusgalli 58, 59
var. *austrojaponensis* 59
var. *hispidula* 59
cruspavonis 59
eruciformis 60
esculenta 58, 59
frumentacea 59
subsp. *utilis* 59
glabrescens 59
oryzoides 58, 59
picta 58, 59
stagnina 59
utilis 59
Ectrosia 44
agrostoides 44
lasioclada 44
leporina 44
Ectrosiopsis 44
lasioclada 44
subaristata 44
Ehrharta 24
calycina 24
diplax 24
erecta 24
stipoides 24
EHRHARTEAE 4, 24
Eleusine 46
coracana 46
filiformis 45
indica 46
Elionurus 92
citreus 92
Elymus 34
repens 34
Enteropogon 49
dolichostachyus 49
unispiceus 3, 49
Entolasia 62
marginata 62
ERAGROSTIDEAE 4, 38
Eragrostis 38
amabilis 44
var. *plumosa* 44
atropioides 38, 41
atrovirens 41
f. *brownii* 41
bahiensis 41
brownii 40, 41
carolinensis 43
cilianensis 39, 42
ciliaris 39, 40, 42
coerulea 43
curvula 40, 42
deflexa 38, 42
depallens 43
diandra 42
dielsii 39, 42
elliottii 40, 41, 42
elongata 40, 42
elytroblephara 44
equitans 44
fosbergii 3, 38, 42
grandis 38, 42
var. *oligantha* 42
var. *polyantha* 42
hawaiiensis 44
hobdyi 44
hosakai 3, 42
japonica 39, 40, 42
lasioclada 44
leptophylla 38, 42
leptostachya 3, 39, 42
marquisensis 45
mauiensis 3, 41, 43
minor 40, 43
molokaiensis 41
monticola 38, 43
niihauensis 44
novocaledonica 43
parviflora 39, 43
paupera 39, 43
pectinacea 40, 43
petersii 43
phleoides 44
pilosa 39, 40, 43
polyantha 42
pooides 43
scabriflora 41, 43
spartinoides 39, 40, 43
subaristata 44
superba 39, 43
tef 38, 40, 43
tenax 43
tenella 39, 44
var. *insularis* 44
tenuifolia 41, 44
thyrsoidea 44
trichodes 39, 44
unioloides 39, 40, 44
variabilis 38, 39, 44

var. *ciliata* 44
wahowensis 44
whitneyi 43
var. *caumii* 43
xerophila 45
Eremochloa 93
ophiuroides 93
Eriachne 78
pallescens 78
ERIACHNEAE 4, 78
Erianthus maximus 80
var. *seemannii* 80
pedicellaris 80
var. *rapensis* 80
Eriochloa 62
procera 62
punctata 62
Eulalia 82
aurea 82
fulva 82
glabrata 82
gracillima 82
spectabilis 83
Eustachys 49
petraea 49

Festuca 25
arundinacea 25
briquetii 26
bromoides 26
fusca 45
hawaiiensis 3, 25
myuros 26
pratensis 25
rubra 25, 26
sandvicensis 27

Garnotia 78
acutigluma 79
cheesemanii 3, 79
var. *rarotongensis* 79
depressa 79
divergens 78, 79
foliosa 79
gracilis 79, 80
linearis 79, 80
mucronata 79
munroana 80
raiateensis 79, 80
rarotongensis 79
sandwicensis 79
solitaria 80
st-johnii 79, 80
stricta 79, 80
var. *marianarum* 80
villosa 79, 80
Gastridium 32
ventricosum 32
Glyceria 28
fluitans 28
subsp. *plicata* 28
notata 28
plicata 28
Greslania circinnata 3
rivularis 3
Gymnothrix complanata 74
macrostachys 74
Gynerium jubatum 36

Hackelochloa 93
granularis 93
Hemarthria 93
altissima 93
compressa 93
Heteropholis cochinchinensis 93
Heteropogon 91
contortus 3, 91
Holcus 30
bicolor 83
cafer 83
halepensis 84
lanatus 30
nitidus 84
pertusus 86
Hordeum 35
brachyantherum 35
leporinum 35
murinum 35
subsp. *leporinum* 35
vulgare 35
Hydropyrum latifolium 24
Hyparrhenia 91
dregeana 91
hirta 91
rufa 91

Imperata 81
conferta 81
cylindrica 3, 81
exaltata 81

Isachne 76
brassii var. *villosa* 78
carolinensis 77
comata 77
confusa 77
var. *purpurascens* 77
dispar 78
distichophylla 77
globosa 77
var. *ciliaris* 78
kunthiana 77, 78
miliacea 77
var. *minutula* 78
minutula 77, 78
pallens 77, 78
ponapensis 77
pulchella 77, 78
purpurascens 77
villosa 77, 78
vitiensis 77, 78
ISACHNEAE 4, 76
Ischaemum 86
aristatum 87
barbatum 86, 87
byrone 3, 87
chordatum 88
digitatum var. *polystachyum* 88
foliosum 87
var. *leiophyllum* 87
indicum 87
intermedium 88
involutum 62
littorale 87
longisetum 87
var. *raulersoniae* 87
lutescens 87
murinum 87
muticum 87, 88
var. *aristuliferum* 88
ophiuroides 93
polystachyum 3, 87, 88
var. *chordatum* 88
var. *hillii* 88
var. *intermedium* 88
rugosum 86, 88
var. *segetum* 88
secundatum 68
segetum 88
setaceum 87, 88
stokesii 87
timorense 87, 88
vitiense 87
Ixophorus 67
unisetus 67

Koeleria 29
glomerata 29
macrantha 29
nitida 29
vestita 29

Lachnagrostis chamissonis 31
filiformis 31
Lagurus cylindricus 81
Lamarckia 26
aurea 26
Leptaspis 23
angustifolia 23
banksii 23
cochleata 23
lanceolata 23
urceolata 23
zeylanica 23
Leptochloa 45
capillacea 45
decipiens 45
subsp. **decipiens** 45
filiformis 45
fusca 45
subsp. **uninervia** 45
marquisensis 3, 45
panicea 45
subsp. **panicea** 45
uninervia 45
virgata 45
xerophila 3, 45
Lepturopetium 51
kuniense 51
marshallense 51
Lepturus 51
acutiglumis 51
cinereus 51
gasparricensis 51
mildbraedianus 51
pilgerianus 51
repens 51
var. *cinereus* 51
var. *maldenensis* 51
var. *occidentalis* 51
var. *palmyrensis* 51

var. *septentrionalis* 51
var. *subulatus* 51
Lolium 26
multiflorum 26
perenne 26
temulentum 26
Lophatherum 35
gracile 35

Megastachya uninervia 45
Megathyrsus maximus 56
Melica latifolia 37
MELICEAE 4, 28
Melinis 68
minutiflora 3, 68
repens 68, 69
Microlaena avenacea 24
stipoides 24
Microstegium 82
glabratum 82
gracillimum 82
pseudeulalia 90
spectabile 82, 83
tenue 82, 83
Milium compressum 65
globosum 77
punctatum 62
Miscanthus 81
floridulus 81
japonicus 81
sinensis 81
Mnesithea 93
laevis 93
Moorochloa eruciformis 61
Muhlenbergia 48
microsperma 48

Nassella cernua 25
Nastus elatus 3
obtusus 3
productus 3
Nemastachys taitensis 82
Neurachne montanum 57
torrida 57
Notodanthonia biannularis 36
semiannularis 36

Ophiurinella micrantha 68
Oplismenus 52
burmannii 52
compositus 52
f. *glabratus* 52
f. *pubescens* 52
var. *patens* 52
var. *setarius* 52
hirtellus 52
subsp. *imbecillis* 52
var. *imbecillis* 52
var. *microphyllus* 52
imbecillis 52
microphyllus 52
oahuaensis 52
patens 52
setarius 52
f. *sterilis* 52
undulatifolius 52
var. *imbecillis* 52
Orthopogon imbecillis 52
Oryza 23
neocaledonica 3, 23, 24
sativa 23, 24
ORYZEAE 4, 23
Oryzopsis 25
miliacea 25
Otatea acuminata 3
aztecorum 3

PANICEAE 4, 52
Panicum 1, 52
abyssinicum 70
accrescens 58
adscendens 71
affine 56
alakaiense 55
amabilis 55
ambiguum 61
americanum 74
annuale 55
antidotale 54
assurgens 57
austrocaledonicum 66
baltodes 55
barbatum 66
beecheyi 3, 53, 54
bifurcatum 57
brizanthum 60
burmannii 52
capillare 54
carteri 55
ciliare 71

cinereum 57
colliei 56
collinum 72
colonum 59
coloratum 54
comae 57
compositum 52
conjugens 55
cookei 55
crusgalli 59
cynodon 53, 55
dactylon 50
decompositum 54, 55
degeneri 55
dimidiatum 68
distachyon 60
distans 67
divaricatissimum 71
ekeanum 56
elegantulum 67
eruciforme 60
esculentum 59
fauriei 3, 53, 55
  var. *carteri* 55
  var. *latius* 55
filiforme 71
flavidum 67
forbesii 55
furtivum 57
gaudichaudii 71
glaucum 74
glumare 61
gossypinum 57
gracile 67
gracilius 55
havaiense 56
heupueo 57
hillebrandianum 53, 55
  var. *gracilius* 55
hirtellum 52
hispidulum 59
hobdyi 55
honokowaiense 56
humidicola 61
imbricatum 55
  f. *minus* 55
  f. *molokaiense* 55
  var. *oreoboloides* 55
infidum 61
infraventale 55
isachnoides 3, 53, 55
  var. *kilohanae* 55
italicum 66
kaalaense 56
kahiliense 55
kahoolawense 57
kanaioense 55
kaonohuaense 57
kauaiense 54
knudsenii 55
kokeense 55
konaense 53, 55
koolauense 3, 53, 56
kukaiwaaense 55
kunthianum 78
lamiatile 55
lanaiense 57
lihauense 54
lineale 3, 54, 56
longivaginatum 54, 56
lustriale 55
luzonense 54, 56
malikoense 55
marginatum 62
maximum 3, 52, 56
  var. *trichoglume* 56
microbachne 72
miliaceum 53, 56
miliiforme 62
minutulum 78
mokuleiaense 54
molle 61
molokaiense 57
montanum 57
monticola 55
moomomiense 55
mosambicense 61
muticum 61
nephelophilum 53, 56
  var. *levius* 56
  var. *rhyacophilum* 57
  var. *tenuifolium* 57
  var. *xerophilum* 57
niihauense 3, 53, 56
ninoleense 55
nubigenum 57
  var. *latius* 55
numaeense 57
oahuaense 52
ooense 55

oreoboloides 55
var. *subimbricatum* 55
oryzoides 59
oxyphyllum 58
palauense 54, 56
pallidefuscum 66
palmifolium 66
paludosum 54, 56
patens 58
patulum 61
pellitoides 57
pellitum 53, 56
pepeopaeense 55
pictum 59
plantagineum 61
polystachion 74
prostratum 61
var. *marquisense* 61
pruriens 72
pseudagrostis 56
pulchellum 78
pumilum 66
radicosum 72
ramosius 3, 53, 57
repens 3, 54, 57
reptans 61
semialatum 59
setarium 52
simplex 57
sphacelatum 67
stenotaphrodes 73
subglabrum 57
subquadriparum 62
sylvanum 55
taitense 61
telmatodes 56
tenuifolium 53, 57
tenuissimum 48
timorense 72
torridum 53, 57
trichoides 53, 57
trigonum 58
uncinulatum 57
verticillatum 67
waikoloaense 55
waimeaense 56
wilburii 55
wiliwilinuiense 77
xerophilum 53, 57
Paspalidium 67
distans 67
elegantulum 67
flavidum 67
Paspalum 62
bicorne 71
cartilagineum 64
var. *biglumaceum* 64
ciliatifolium 64
commersonii 64
conjugatum 3, 63
dilatatum 63
distichum 3, 63, 64
distichum auct. 65
fimbriatum 63, 64
fissifolium 65
forsterianum 63, 64
fuscescens 71
heteranthum 71
longiflorum 72
longifolium 63, 64
macrophyllum 63, 64
malacophyllum 62, 64
moratii 64
notatum 63, 64
orbiculare 64
var. *otobedii* 64
paniculatum 63, 64
paschale 65
scoparium var. *oligostachyum* 65
scrobiculatum 63, 64
setaceum 63, 64
var. *ciliatifolium* 64
thunbergii 63, 65
undulatum 64
urvillei 63, 65
vaginatum 63, 65
venustum 64
virgatum 63, 65
Pennisetum 73
americanum 74
articulare 74
caliculatum 76
clandestinum 3, 73, 74
complanatum 73, 74
glaucum 73, 74
henryanum 74
var. *longisetum* 74
var. *pluristylum* 74
identicum 74
macrostachyum 74

marquisense 3, 76
orientale 73, 74
polystachion 3, 73, 74
f. *viviparum* 74
purpureum 3, 73, 75
setaceum 3, 73, 75
setosum 74
simeonis 74
var. *intermedium* 74
var. *pedicellatum* 74
var. *purpureum* 74
triflorum 74
villosum 73, 75
Phalaris 30
aquatica 30
canariensis 30
hispida 91
minor 30
paradoxa 30
semiverticillata 33
setacea 75
stenoptera 30
tuberosa 30
var. *stenoptera* 30
zizanioides 84
PHAREAE 4, 23
Pharus urceolatus 23
Phleum 33
cochinchinense 93
indicum 82
pratense 33
Phragmites 37
australis 37
communis 37
karka 37
vallatorius 37
Phyllostachys aurea 3
nigra 3
Piptatherum miliaceum 25
Poa 27
amabilis 44
annua 27
arachnifera 27
atrovirens 41
brownii 41
cilianensis 42
ciliaris 42
compressa 27
curvula 42
decipiens 45
diandra 42
elongata 42
japonica 42
leptostachya 42
longeradiata 27
malabarica 45
mannii 27
monticola 43
panicea 45
parviflora 43
pectinacea 43
pilosa 43
plumosa 44
pratensis 27
sandvicensis 3, 27
siphonoglossa 3, 27, 28
tef 43
tenella 44
tenuifolia 44
trichodes 44
unioloides 44
variabilis 44
POEAE 4, 25
Pogonatherum 82
crinitum 82
paniceum 82
Pollinia gracillima 82
spectabilis 83
tenuis 83
Polypogon 32
fugax 32
interruptus 32, 33
monspeliensis 32, 33
viridis 32, 33
Polytoca macrophylla 94
Polytrias 82
amaura 82
indica 82

Racemobambos holttumii 3
Rhaphis aciculata 84
arundinacea 83
Rhynchelytrum repens 69
roseum 69
Rottboellia 93
altissima 93
cochinchinensis 93
coelorachis 93
compressa 93
exaltata 93

laevis 93
repens 51
Rytidosperma 35
biannulare 35, 36
paschale 36
pilosum 35, 36
semiannulare 36

Saccharum 80
confertum 81
edule 81
floridulum 81
fulvum 82
insulare 81
maximum 80
officinarum 80
paniceum 82
pedicellare 80
repens 69
robustum 81
spontaneum 80, 81
var. *edule* 81
var. *insulare* 81
Sacciolepis 58
indica 58
Schedonorus arundinaceus 25
pratensis 25
Schizachyrium 90
brevifolium 90
condensatum 90
fragile 90
obliquiberbe 90
paradoxum 90
pseudeulalia 90
scoparium 90, 91
Schizostachyum glaucifolium 3
tessellatum 3
Scrotochloa 23
urceolata 23
Setaria 65
austrocaledonica 66
barbata 65, 66
biflora 68
elegantula 67
geniculata 66
glauca 74
glauca auct. 67
gracilis 66
italica 66
jaffrei 3, 66
lutescens 67
pallidefusca 66
palmifolia 65, 66
parviflora 66
pumila 66
sphacelata 66, 67
verticillata 66, 67
Sorghum 83
arundinaceum 83, 84
bicolor 83, 84
var. *cafer* 83
var. *obovatum* 83
var. *rotundulum* 83
var. *subglabrescens* 83
var. *transiens* 84
×drummondii 3, 83, 84
halepense 3, 83, 84
f. *muticum* 84
var. *propinquum* 84
miliiforme var. *rotundulum* 83
nitidum 83, 84
propinquum 83, 84
sudanense 84
verticilliflorum 83
vulgare 83
var. *sudanense* 84
Spinifex 76
hirsutus 76
littoreus 76
sericeus 76
Spodiopogon byronis 87
chordatus 88
Sporobolus 46
africanus 47, 48
diandrus 46, 47
elongatus 47
farinosus 46, 47
fertilis 47, 48
indicus 47, 48
var. *africanus* 47
var. *diandrus* 47
var. *fertilis* 47
indicus auct. 47
jacquemontii 48
piliferus 46, 47
pyramidalis 47, 48
pyramidatus 46, 48
tenuissimus 46, 48
virginicus 46, 48
Stegosia cochinchinensis 93

Stenotaphrum 68
dimidiatum 68
micranthum 68
secundatum 68
subulatum 68
Stipa 1, 25
arguens 92
cernua 25
horridula 3, 25
micrantha 32
scabra 3, 25
STIPEAE 4, 25
Syntherisma helleri 73
mariannensis 72
mezii 72
pelagica 73
pruriens 72
robinsonii 71
stenotaphrodes 73
stricta 71

Themeda 91
arguens 91, 92
australis 92
ciliata 92
gigantea 92
var. *intermedia* 92
intermedia 92
quadrivalvis 91, 92
triandra 92
villosa 92
Thuarea 62
involuta 62
sarmentosa 62
Thysanolaena 37
latifolia 37
THYSANOLAENEAE 4, 37
Trachypogon rufus 91
Tragus 50
australianus 50
berteronianus 50
racemosus 50
Trichachne insularis 72
Trichochloa microsperma 48
Tricholaena rosea 69
Tripsacum 94
andersonii 94
hermaphroditum 76
latifolium 94
laxum 94
Trisetum 29
flavescens 29
glomeratum 29
inaequale 3, 29
TRITICEAE 4, 34
Triticum 34
aestivum 34
repens 34
sativum 34

Uniola spicata 38
Urachne acutigluma 79
Urochloa 59
ambigua 61
brizantha 60
decumbens 60
distachya 60
eruciformis 60
glumaris 59, 61
humidicola 60, 61
maxima 56
mollis 60, 61
mosambicensis 59, 61
mutica 60, 61
paspaloides 61
plantaginea 60, 61
reptans 60, 61
subquadripara 60, 62
uniseta 67

Vetiveria zizanioides 84
Vilfa pilifera 47
Vulpia 26
bromoides 26
myuros 26

Zea 94
mays 94
Zizania 24
latifolia 24
Zoysia 50
japonica 50, 51
matrella 50
subsp. *japonica* 51
var. *japonica* 51
var. *pacifica* 51
pacifica 51
pungens 51
tenuifolia 51

Lightning Source UK Ltd.
Milton Keynes UK
16 June 2010

155632UK00001BA/35/P